AA008669

1st International Kaidalov Workshop on Phenomenology of High Energy Particle Physics 2013

Journal of Physics: Conference Series Volume 607

Moscow, Russia
21-25 July 2013

ISBN: 978-1-5108-0751-8
ISSN: 1742-6588

Printed from e-media with permission by:

Curran Associates, Inc.
57 Morehouse Lane
Red Hook, NY 12571

Printed by Curran Associates, Inc. (2015)

For permission requests, please contact the Institute of Physics
at the address below.

Institute of Physics
Dirac House, Temple Back
Bristol BS1 6BE UK

Phone: 44 1 17 929 7481
Fax: 44 1 17 920 0979

techtracking@iop.org

Additional copies of this publication are available from:

Curran Associates, Inc.
57 Morehouse Lane
Red Hook, NY 12571 USA
Phone: 845-758-0400
Fax: 845-758-2634
Email: curran@proceedings.com
Web: www.proceedings.com

Table of contents

Volume 607

1st International Kaidalov Workshop on the Phenomenology of High Energy Particle Physics
21–25 July 2013, Moscow, Russia

Accepted papers received: 27 March 2015
Published online: 14 May 2015

Preface

Papers

I. Phenomenology of particle production at LHC and Regge theory models

II. Phenomenology of new physics in astroparticle and collider experiments

III. Phenomenology of strong interactions with the computer experiments

The 1st International Workshop on Phenomenology of High Energy Particle Interactions, devoted to the memory of Alexei Kaidalov, was organized in Moscow on 21-25 July 2013. The aim of workshop was to gather physicists who have worked with A. Kaidalov, as well as to collect the subjects of his articles that are important for further investigations. We took a broad area of topics in order to initiate an active intercrossing between experimentalists and theorists in the field of high energy particle interactions at collider experiments, as in the space observations. We hope the Workshop can gather more young physicists from our country and worldwide who are interested in understanding the high energy physics and also ready to participate in the real active work related to the problems of up-to-date particle physics.

The topics discussed were:

Strong Interactions Phenomenology.

New physics in astroparticle and collider physics.

Computer Experiments in Particle Physics.

International Organizing Committee of Workshop:

Valentin Zakharov (ITEP, Russia)

Mikhail Danilov (ITEP, Russia)

Mikhail Polikarpov (ITEP, Russia)

Mikhail Vysotski (ITEP, Russia)

Dmitri Kharseev (Uni. of Stony Brook, USA)

Vladimir Nikitin (JINR, Dubna, Russia)

Dmitri Kazakov (JINR, Dubna, Russia)

Larissa Bravina (Uni. of Oslo, Norway)

Lidia Smirnova (Moscow Uni., Russia)

Caros Merino (Uni. of Santiago de Compostela, Spain)

Alexei Kurepin (INR, Russia)

Oleg Dalkarov (LPI, Russia)

Olga Piskounova (LPI, Russia)

The Organizing Committee is thankful to the agencies and institutions that supported our Workshop:

Russian Foundation for Basic Research (grant N13-02-06091), Noncommercial Foundation "Dynasty", Institute of Theoretical and Experimental Physics and the administrative department of P. N. Lebedev Physical Institute of RAS. We express our gratitude to administration of Business school SKOLKOVO for the excellent management of the lectures, coffee breaks and receptions.

Personally, Prof. A.B. Kurepin (INR, Moscow), Prof. D.I. Kazakov (JINR, Dubna) and Dr. S.P. Baranov helped a great deal. Just before the meeting Mikhail Polikarpov, who was the active organizer of Workshop, unexpectedly passed away and our duty is to publish here the memories of his students and coworkers.

On behalf of Organizing Committee

Dr. Olga Piskounova

Journal of Physics: Conference Series **607** (2015) 011002 doi:10.1088/1742-6596/607/1/011002

Journal of Physics: Conference Series **607** (2015) 011003 doi:10.1088/1742-6596/607/1/011003

Peer review statement

All papers published in this volume of *Journal of Physics: Conference Series* have been peer reviewed through processes administered by the proceedings Editors. Reviews were conducted by expert referees to the professional and scientific standards expected of a proceedings journal published by IOP Publishing.

Journal of Physics: Conference Series **607** (2015) 012001 doi:10.1088/1742-6596/607/1/012001

Tribute to Alexei Borisovich Kaidalov

C. Merino

Departamento de Física de Partículas, Facultade de Física, and Instituto Galego de Física de Altas Enerxías (IGFAE), Universidade de Santiago de Compostela, 15782 Santiago de Compostela, Galiza-Spain

E-mail: `merino@fpaxp1.usc.es`

Abstract. The author wrote these lines to recollect some personal memories of Alexei Borisovich Kaidalov and of the collaboration the author carried out with him under his guidance for almost 20 years. By doing so the author has once again confirmed the realization of the deep influence that Alexei Borisovich and his contribution to physics had and still has in the spirit of the author, and in the formation of several generations of physicists, both from Russia and from abroad.

Alexei Borisovich Kaidalov
(20.07.1940 - 25.07.2010)

I first personally met Alexei Borisovisch Kaidalov (for some biographical reference on Alexei see reference [1]) in the late 80's of last century, during the preparation of my Ph.D. thesis under the supervision of Alfons Capella in the Laboratoire de Physique Théorique et des Hautes Énergies (LPTHE) of the Université de Paris-Sud, Paris XI (France), when Alexei visited the laboratory to collaborate with Alfons and Jean Tran Thanh Van.

I remember that from the very first moment Alexei made a big impression on me, impression reinforced by his patience to invest time in discussing physics with the young doctoral student I was at the time.

After passing my Ph.D. in April 1991, I had to go for my military service in the Spanish Navy for one year, and shortly after finishing my service, when I was wondering which would be the best way of tackling my postdoc activity, in July 1992 Alexei attended, with Konstantin G. Boreskov and with the great Karen A. Ter-Martirosyan, all from the Institute of Theoretical and Experimental Physics (ITEP) of Moscow, the XXII International Symposium on Multiparticle Dynamics, organized in my local University of Santiago de Compostela by the head of the Particle Physics Department, and also advisor of my Ph.D. thesis, Professor Carlos Pajares. The fact of having such a great physicists from a so prestigious institution as ITEP visiting our university was encouraging for the young doctorates in Santiago de Composatela at the start of our research careers.

During this stay in Santiago de Compostela, Karen, Alexei and Konstantin invited me to visit ITEP, where I spent one month in January-February 1993. Apart from the big experience of being in such an important place for the history of theoretical physics during the twentieth century, and from getting a closer image of extraordinary figures, like Isaak Yakovlevich Pomeranchuk, of whom Alexei was the last Ph.D. student, I was friendly introduced to the Russian culture, and treated to enjoy the canonical way of sharing zakuska and vodka.

Later, in October that year 1993, I started a postdoctoral two-year stay at LPTHE in Paris, in the frame of the Human Capital and Mobility Program of the European Union, and Alexei came there to spend several months at the beginning of 1994. The presence of Alexei made an immediate impact on the research activity of the group led by Alfons Capella, and that soon translated into the writing of several papers devoted to the description of low-x physics, papers that I cannot resist the temptation of citing here. In the first of these papers [2], and following the conviction by Alexei that a common description of both the soft and the hard regimes of strong interactions was possible, a phenomenological model based on Regge Theory for the initial condition of $F_2(x, Q^2)$ in the perturbative QCD evolution was presented. In what it was called the CKMT model, we considered the following behaviour of the low-x F_2 structure function:

$$F_2(x, Q^2) \sim x^{-\Delta(Q^2)} \; ; \; \Delta(Q^2) = \Delta_0 \cdot \left(1 + \frac{2Q^2}{Q^2 + d}\right) , \tag{1}$$

where Δ_0 and d are free parameters. From the fit to experimental data, the value $\Delta(Q^2 = 0) = \Delta_0 \sim 0.08$ was obtained, what corresponds to $\Delta(Q^2 \to \infty) \sim 0.24$ (bare Pomeron). According to large N-expansion, we assumed that the Pomeron determines the small-x behaviour of sea and gluons, while secondary reggeons (ρ,ω,f,A_2), with intercept $\alpha_R \sim 0.4 - 0.6$, determine the behaviour of valence quarks. Thus, the following parametrization of the structure functions in the region of small and moderate Q^2 ($0 \leq Q^2 \leq 5\ GeV^2$), was implemented:

$$\begin{aligned} F_2(x, Q^2) &= A \cdot x^{-\Delta(Q^2)}(1-x)^{n(Q^2)+4} \cdot \left(\frac{Q^2}{Q^2+a}\right)^{1+\Delta(Q^2)} \\ &+ B \cdot x^{1-\alpha_R}(1-x)^{n(Q^2)} \cdot \left(\frac{Q^2}{Q^2+b}\right)^{\alpha_R} . \end{aligned} \tag{2}$$

The first term in equation 2 corresponds to the Pomeron contribution with a $x \to 0$ behaviour given by equation 1. The second term corresponds to the secondary reggeon contribution with $x \to 0$ behaviour determined by the secondary reggeon intercept α_R. The behaviour at $x \to 1$ is given by the second factor, with $n(Q^2)$ parametrized as

$$n(Q^2) = \frac{3}{2} \cdot \left(1 + \frac{Q^2}{Q^2 + c}\right) . \quad (3)$$

in such a way that, at $Q^2 = 0$, one recovers the same power $(1-x)^{1.5}$ as in the Dual Parton Model (controlled by Regge intercepts, $n(0) = \alpha_R(0) - 2\alpha_N(0) \sim \frac{3}{2}$), and at $Q^2 \to \infty$ it goes as $(1-x)^3$, as given by dimensional counting rules. The last factor in the two terms of equation 2 is required by equation 1 in order to connect with real photons. By using equation 1, one gets:

$$\sigma_{\gamma p}^{tot}(\nu) = 4\pi^2\alpha_{EM}\left(Aa^{-1-\Delta_0}(2m\nu)^{\Delta_0} + Bb^{-\alpha_R}(2m\nu)^{\alpha_R - 1}\right) . \quad (4)$$

The parameters into the model were determined from a joint fit of the $\sigma_{\gamma p}^{tot}$ data and of the NMC data on the proton structure function $F_2(x, Q^2)$ in the region $1 \le Q^2 \le 5\ GeV^2$, and the contribution by u and d valence quarks appears separated as:

$$B \cdot (1-x)^{n(Q^2)} \to B_u \cdot (1-x)^{n(Q^2)} + B_d \cdot (1-x)^{n(Q^2)+1} . \quad (5)$$

The condition of having two valence u-quarks (charge 2/3) and one valence d-quark (−1/3), determines the values of B_u and B_d. A good fit to the experimental data was obtained up to $Q^2 \sim 10\ GeV^2$, even before plugging the perturbative evolution in.

Alexei Kaidalov was perhaps the best expert on diffraction in the world, and we devoted a second paper [3] to diffraction, in which we considered the diffration dissociation of virtual photons in the framework of conventional Regge theory. In that paper we showed that the HERA data on large rapidity gap events were successfully described in terms of the Pomeron structure function, and we obtained the size of the shadowing corrections at low x and large Q^2.

During these years in Paris, and for many years after, I met Alexei in many conferences and meetings, in particular in Les Rencontres de Moriond, organized every March by Jean Tran Van Thanh, either in Les Arcs or Méribel (France), either in La Thuile (Italy), and where Alexei's skiing exploits almost matched up with the brilliance of his talks. Actually, I was always amazed on how Alexei could, after skiing and devoting time to discuss physics with the many participants who asked him for his advice, prepare such a precise and complete summary talks, that actually gave you the clear and paramount image of the physics activity presented during the conference. In January-February 1998 I stayed at ITEP a second time for 1 month, and the welcome by all the colleagues there, led by Alexei, was again friendly and really fruitful to me.

I have collaborated with Alexei for almost twenty years, on structure functions, diffractive dissociation, hard diffraction, strange baryon production, nuclear structure functions, etc, and I share with him the authorship of fifteen entries (some of them also with others) in hep-Spires, the last one from February 2011. I'm truly aware that this represents a very minimal part of the titanic work and the core contribution Alexei has made to particle physics, but to cite our common papers, to be the M after the K of the CKMT model, and to think of the work we shared and of everything I have learned from him along the years makes me always feel deeply proud.

One concise scheme [1] of Alexei's main scientific goals, based on a document prepared by Alexei himself several years before his death, and in which many important results are for sure omitted, could be the following:

Journal of Physics: Conference Series **607** (2015) 012001 doi:10.1088/1742-6596/607/1/012001

1. The consideration of non-lepton and radiative decays of hyperons, the analysis of the amplitudes of these decays and the calculation of the low boundaries for the decay probabilities.
2. The identification for moving reggeon cuts signatures in experimental data. This meant a very important progress at the time, since the standard thinking was that the Regge pole contribution is enough for the description of the experimental data. Alexei also gave the first model-independent estimate, obtained from particle-reggeon unitarity, of the two-reggeon cut contribution.
3. The reformulation of the one-pion-exchange model on the base of Reggeon approach, and the consequent quantitative description of a large number of experimental data on $pp \to pn\pi^+$, $pp \to pp\pi^+\pi^-$, etc., at intermediate energies ($1 - 10\ GeV/c$), as well as the description of the inclusive processes at high energies. All this resulted in a significant progress in the understanding of the nature of secondary production at these energies.
4. The analysis of high energy diffraction dissociation processes. The investigation of low mass diffractive jets allowed the implementation of the so-called quasi-eikonal model for elastic high energy hadron-hadron scattering. The parameters of the triple-Reggeon interactions which play a fundamental role in the processes of high energy interactions were extracted from the experimental data on large mass diffractive jet production. The main results of this activity were published in a brilliant and comprehensive review, still fully standing today, published in Physics Reports [4].
5. The development of the dispersion sum rules for the Reggeon-hadron scattering that allows to predict some high-spin resonances.
6. The formulation of Quark-Gluon String Model (QGSM) for multiparticle production at high and superhigh energies that allowed the quantitative description of a lot of experimental data in hadron-hadron and hadron-nucleus collisions. The QGSM is an analytical, not Monte Carlo, model based on the topological 1/N expansion of QCD amplitudes, on Regge-asymptotics of quark and diquark distributions in the colliding hadrons, and on similar Regge-asymptotics for the fragmentation functions for secondary particle production. The QGSM predictions have been successfully used for the description of the production of secondary particles at collider energies, and they are now confirmed by the corresponding productions at LHC. The QGSM explains, in particular, the baryon charge diffusion at large distances in rapidity space. The QGSM predictions have been also widely used in cosmic ray physics.
7. The analysis of particle production in heavy ion collisions, in which it was shown that the inelastic nuclear screening effects cannot practically change the integral interaction cross section, but they very significantly decrease the inclusive densities of secondaries produced in the central (midrapidity) region.
8. The analysis of heavy quarks and lepton pair production on nucleon and nuclear targets.

Throughout his unusually fruitful career, Alexei Kaidalov wrote more than 300 scientific papers, 16 of them cited more than 100 times (for a far from complete list of publications by Alexei, see reference [1]).

Alexei mantained scientific contacts with many physicists, both theorists and experimentalists, working in high energy and cosmic ray physics, and, thus, he was actively collaborating for many years with the groups of Alfons Capella at the Laboratoire de Physique Théorique et Hautes Énergies (LPTHE), then renamed as LPT, in Orsay (France), and with the group of Alan D. Martin at the University of Durham (UK), and with other groups of different countries. In particular, Alexei was in close connection with several experiments at CERN, where he was a member of the ALICE Collaboration.

Alexei was a respected and admired magister by colleagues and friends from all over the world for his many and very important, often seminal, original contributions to the field of particle physics, but here I would specially underline Alexei's genuine generosity in passing to younger physicists, through clear, and natural explanations, his deep knowledge of the fundamental principles and rules at work under the physics processes.

Aliosha passed away in Moscow, on July 25th, 2010, and we all, colleagues, friends, and students, who heartily miss him, will for ever enjoy and profit from his knowledge and wisdom.

To collaborate with Alexei has ment to me to approach greatness in physics, and a real example of integrity in life.

On top of his dedication to physics and science, Aliosha was a friendly and cheerful person, cultivated, and open-minded, and proud of the Russian culture. He was devoted to his wife, Natascha, and to his family. I want to finish this text by expressing my personal tribute and thoughts of affection to Natasha and to all Alexei's family.

Acknowledgements

I want to thank Konstantin G. Boreskov for providing me with original material by Alexei used in the preparation of this text, and for sharing with me many of his close collaboration and friendship experiences with Alexei. I also thank Gevorg H. Arakelyan and Yuli M. Shabelski for his help and advice in writing this contribution, and for translating to English some original Russian documents. Finally I express my congratulations to Olga Piskunova and all organizers for putting in place this workshop, aimed at the celebration of the memory of Alexei Borisovich Kaidalov, and at insuring the passing of the torch projecting the light of Alexei's figure and work to new generations of physicists all over the world.

References

[1] *In Memoriam: Alexei Borisovich Kaidalov (1940–2010), Proceedings of the Low-x Meeting 2011*, editors C. Merino, G. Parente, and C. Royon (Meubook, ISBN 978-84-940829-9-3), pages 1–7, Hotel Monumento San Francisco, Santiago de Compostela, Galiza, Spain, June 3rd–7th, 2011.

[2] *Structure Functions and Low-x Physics*, by A. Capella, A. Kaidalov, C. Merino, and J. Tran Thanh Van, *Phys. Lett. B***333** (1994) 358 and *hep-ph*/9405338.

[3] *Diffractive Dissociation in Deep Inelastic Scatering at HERA*, by A. Capella, A. Kaidalov, C. Merino, and J. Tran Thanh Van, *Phys. Lett. B***343** (1995) 403 and *hep-ph*/9407372.

[4] *Diffractive Production Mechanisms*, by A.B. Kaidalov, *Phys. Rep.* **50** (1979) 157.

Lessons from LHC elastic and diffractive data

Alan D. Martin
Institute for Particle Physics Phenomenology, University of Durham, Durham, DH1 3LE
E-mail: a.d.martin@durham.ac.uk

Abstract.
We discuss a model which gives a 'global' description of the wide variety of high-energy elastic and diffractive data that are presently available, particularly from the LHC experiments. The model is based on only one pomeron pole, but includes multi-pomeron interactions. Significantly, the LHC measurements require that the model includes the transverse momentum dependence of the intermediate partons as a function of their rapidity, which results in a rapidity (or energy) dependence of the multi-pomeron vertices.

1. Introduction

Since this Workshop is in memory of Aliosha Kaidalov, it is appropriate to list some of his pioneering contributions to the understanding of Diffractive processes in high energy hadron interactions. He was the first to evaluate the effects of low-mass diffractive dissociation, namely processes of the form $pp \to p + N^*$ [1]. Next, Kaidalov et al. [2] were the first to perform a triple-Regge analysis, which underlies the description high-mass dissociation. Together with Ter-Martirosyan, Aliosha proposed [3] that multiple hadron production at high energies was the result of the creation and breaking of quark-gluon strings in hadron-hadron collisions. It is shown that this approach, together with the general results of supercritical Pomeron theory ($\alpha_P(0) > 1$), yields a natural description of all high-energy multiple production data. Again with collaborators, Kaidalov et al. [4, 5] pioneered a model for soft high energy interactions, which included multi-pomeron diagrams in a global description of the data, for the first time. These contributions underlie the models developed for understanding diffractive processes to this day: here we may mention the Durham [6, 7] and Tel-Aviv [8, 9] models, the work by Ostapchenko [10] and by Poghosyan and Kaidalov [11].

Basically the models describe high-energy diffractive processes by pomeron exchange within the framework of Reggeon Field Theory (RFT) [12]. Elastic and low-mass diffractive dissociation are described in terms of a multi-channel eikonal and high-mass diffraction in terms of diagrams with multi-pomeron couplings, the simplest of which is the triple-pomeron diagram.

Recently a wide variety of measurements of diffractive processes has been obtained by experiments at the LHC. Attempts at a 'global' simultaneous description of these data show differences with the expectations of conventional Regge Theory. Here, we explain how, in Refs. [13, 14], it has been possible to modify the classic RFT in a physically-motivated way so as to accommodate the tendencies observed at the LHC. The crucial modification is to include the transverse momentum dependence of the intermediate partons on rapidity (or energy). We may call this the $k_t(s)$ effect. But, first, we list the high-energy diffractive data that exist at present. Then we outline some of the problems presented by the data.

2. The high-energy diffractive data

Quite a wide variety of diffractive measurements have been made at the LHC. At the moment, data for diffractive processes are available at 7 TeV. The most detailed data come from the TOTEM collaboration. TOTEM have measured
(i) the total and elastic cross sections (in a wide t interval including the dip region) [15, 16];
(ii) the cross section of low-mass ($M_X < 3.4$ GeV) proton dissociation [17];
(iii) double dissociation ($pp \to X_1 + X_2$) [18]; and
(iv) made preliminary measurements of high-mass single proton dissociation, $\sigma_{\rm SD}$, integrated over three intervals of M_X: namely (3.4, 8); (8, 350); (350, 1100) GeV [19].
In addition, we have the inelastic cross sections and the cross sections of events with a Large Rapidity Gap (LRG) measured by the ATLAS [20], CMS [21] and ALICE [22] collaborations. Moreover, we have valuable data on elastic and proton dissociation from experiments at the Tevatron [23, 24, 25].

3. Potential puzzles in the diffractive data sets

Following [13, 14], we list several potential puzzles that seem to exist in the data.

(a) First, we note that the total cross section in the Tevatron – LHC energy interval starts to grow *faster*, not more slowly, than below the Tevatron energy, and, secondly, the slope of the effective pomeron trajectory, $\alpha'_{\rm eff}$, increases [26]. In particular, Donnachie-Landshoff-type fit [27] predicts $\sigma_{\rm tot} = 90.7$ mb at $\sqrt{s} = 7$ TeV, while TOTEM observes 98.6±2.2 mb [16]. Moreover, the elastic slope was measured at the Tevatron ($\sqrt{s} = 1.8$ TeV) to be $B_{\rm el} = 16.3 \pm 0.3$ GeV^{-2} by the E710 experiment [23] and to be $B_{\rm el} = 16.98 \pm 0.25$ GeV^{-2} by the CDF group [24]. Even starting from the CDF result, and using the $\alpha'_P = 0.25$ GeV^{-2}, we expect $B_{\rm el} = 16.98 + 4 \times 0.25 \times \ln(7/1.8) = 18.34$ GeV^{-2} at 7 TeV, while TOTEM finds 19.9 ± 0.3 GeV^{-2} [16]. Also, in the relatively low $|t| < 0.3 - 0.4$ GeV^2 region we do not see the expected increase of t-slope as $|t|$ increases.

(b) The experimental information on low-mass dissociation is a puzzle in that the cross section $\sigma_{\rm D}^{{\rm low}M_X}$ goes from about $2-3$ mb at the CERN-ISR energy of 62.5 GeV to only 2.6 ± 2.2 mb at 7 TeV at the LHC [17]. Thus $\sigma_{\rm D}^{{\rm low}M_X}$ is about 30% of $\sigma_{\rm el}$ at 62.5 GeV and only about 10% at 7 TeV, whereas we would expect these percentages to be about the same for single pomeron exchange.

(c) An analogous problem occurs for high-mass dissociation where the cross section $d\sigma_{\rm SD}/d\ln\xi$ in the first (3.4 to 8 GeV) M_X interval is more than twice larger than that in the central interval. Of course, according to the triple-Regge formula, a pomeron intercept $\alpha_P(0) > 1$ leads to an increase of the cross section when ξ decreases, but by the same argument we have to observe a larger cross section at the LHC than at the Tevatron, for the same value of M_X, contrary to the data.

(d) The 'factorisation' relation between the observed elastic, single and double proton dissociation cross sections is intriguing, and appears not easy to explain.

In Refs. [13, 14] it is claimed that a global description, including the $k_t(s)$ effect, is able to account for all these puzzles in the data. This will be the subject of Section 7.

4. Eikonal approach to elastic scattering

To establish notation, we start by considering just elastic proton-proton scattering. In the simplest case, the high-energy elastic scattering amplitude, $T_{\rm el}$, (and correspondingly the total cross section) is parametrized by single pomeron exchange. In practice, the pre-LHC data require the trajectory of this effective (soft) pomeron to be

$$\alpha_P(t) = 1 + \Delta + \alpha'_P t \; , \tag{1}$$

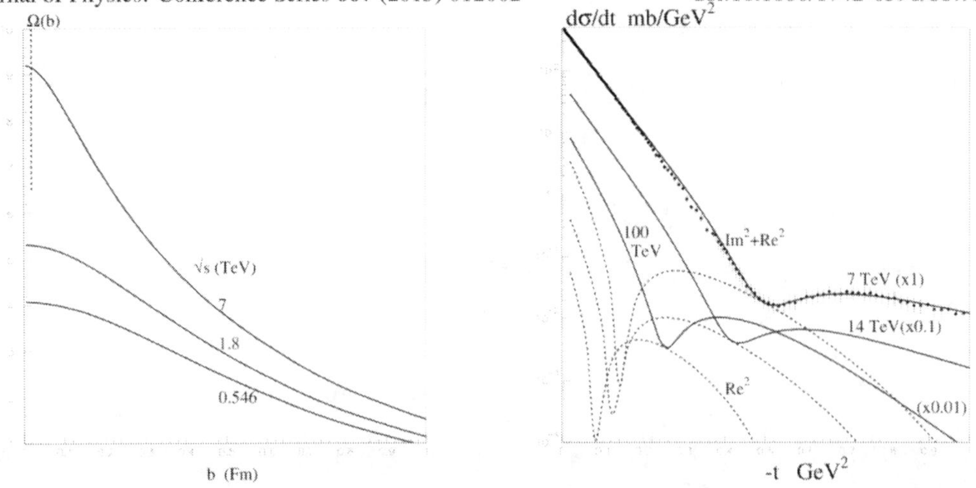

Figure 1. (a) The proton opacity $\Omega(b)$ determined directly from the $d\sigma_{\rm el}/dt$ data at 546 GeV [28, 29, 30], 1.8 TeV [23, 24] and 7 TeV [31] data – the uncertainty on the LHC value at $b = 0$ is indicated by a dashed line; (b) the continuous curves are the total contribution to the elastic cross section, and the dashed curves are the contribution of the real part of the amplitude, at three collider energies: 7, 14 and 100 TeV – the TOTEM data at 7 TeV are shown. The figures are taken from [14].

with $\Delta \simeq 0.08$ and $\alpha'_P \simeq 0.25$ GeV^{-2} [27].

However, already two-particle s-channel unitarity generates a series of the non-enhanced multi-pomeron diagrams leading to the eikonal approximation. The unitarity relaion is of the form

$$2\ {\rm Im}\ T_{\rm el}(b) = |T_{\rm el}(b)|^2 + G_{\rm inel}(b)\ , \tag{2}$$

where G is the sum over all the inelastic intermediate states. The solution gives an elastic amplitude,

$$T_{\rm el}(b) \ = \ i(1 - e^{-\Omega/2})\ , \tag{3}$$

where one pomeron exchange describes the opacity $\Omega(s, b)$, which depends on the square of the incoming energy, s, and the impact parameter, b. Thus we have

$$\sigma_{\rm tot}(s,b) \ = \ 2(1-{\rm e}^{-\Omega/2}), \tag{4}$$

$$\sigma_{\rm el}(s,b) \ = \ (1-{\rm e}^{-\Omega/2})^2, \tag{5}$$

$$\sigma_{\rm inel}(s,b) \ = \ 1-{\rm e}^{-\Omega}, \tag{6}$$

Note, from (6), that

$$S^2(b) \equiv e^{-\Omega} \tag{7}$$

is the probability that no inelastic interaction occurs at impact parameter b. This observation will enable us to calculate the probability that large rapidity gaps survive soft rescattering. In terms of the opacity the elastic cross section takes the form

$$\frac{d\sigma_{\rm el}}{dt} = \frac{1}{4\pi}\left|\int d^2b\ e^{i\vec{q_t}\cdot\vec{b}}(1-e^{-\Omega(b)/2})\right|^2 = \frac{1}{2}\left|\int bdb\ J_0(q_tb)(1-e^{-\Omega(b)/2})\right|^2 \tag{8}$$

where $q_t = \sqrt{|t|}$.

The b dependence of $T_{\rm el}$ can be obtained from the experimental data by taking the Fourier transform data [32]

$${\rm Im}T_{\rm el}(b) \ = \ \int\sqrt{\frac{d\sigma_{\rm el}}{dt}\frac{16\pi}{1+\rho^2}}\ J_0(q_tb)\ \frac{q_tdq_t}{4\pi}, \tag{9}$$

where the square root represents $\text{Im}T_{\text{el}}(q_t)$, with $\rho \equiv \text{Re}T_{\text{el}}/\text{Im}T_{\text{el}}$. In this way, we first determine $\text{Im}T_{\text{el}}$ from the data for $d\sigma_{\text{el}}/dt$, and then calculate $\Omega(b)$ using (3), assuming that ρ is small. In fact, the assumption $\rho^2 \ll 1$ is actually well justified, except in the diffractive dip region; see the discussion below (15).

The results obtained via (9) are shown in Fig. 1(a), where we compare $\Omega(b)$ obtained from elastic differential cross section data at Sp$\bar{p}$S [28, 29, 30], Tevatron [23, 24] and LHC [31] energies. At the lower two energies the $\Omega(b)$ distributions have approximately Gaussian form, whereas at the LHC energy we observe a growth of Ω at small b. The growth reflects the fact that the TOTEM data indicate that we have almost total saturation at $b = 0$. Note that, according to (3), the value of $\text{Im}T(b = 0) \to 1$ corresponds to $\Omega \to \infty$. Since actually we do not reach exact saturation, the proton opacity at $b = 0$ is not ∞, but just large numerically. Clearly, in this region of b, the uncertainty on the value of Ω is large as well, see Fig. 1(a).

5. Inclusion low-mass dissociation of the proton

To describe proton dissociation into *low-mass* states we follow the Good-Walker approach [33], and introduce so-called diffractive (or GW) eigenstates, $|\phi_i\rangle$ with $i = 1, n$, that diagonalize the T-matrix. The incoming 'beam' and 'target' proton wave functions are written as superpositions of the diffractive eigenstates

$$|p\rangle_{\text{beam}} \;=\; \sum a_i|\phi_i\rangle, \qquad |p\rangle_{\text{target}} \;=\; \sum a_k|\phi_k\rangle. \tag{10}$$

It is sufficient to use two diffractive eigenstates, $i, k = 1, 2$. In terms of this 2-channel eikonal model, we have

$$\sigma_{\text{el}} \;=\; \int d^2b \left| \sum_{i,k} |a_i|^2|a_k|^2 \; (1 - e^{-\Omega_{ik}(b)/2}) \right|^2, \tag{11}$$

$$\sigma_{\text{tot}} \;=\; 2\int d^2b \; \sum_{i,k} |a_i|^2|a_k|^2 \; (1 - e^{-\Omega_{ik}(b)/2}) \tag{12}$$

and the 'total' low-mass diffractive cross section

$$\sigma_{\text{el+SD+DD}} \;=\; \int d^2b \; \sum_{i,k} |a_i|^2|a_k|^2 \; \left|(1 - e^{-\Omega_{ik}(b)/2})\right|^2, \tag{13}$$

where SD includes the single dissociation of both protons, DD denotes double dissociation, and where the opacity $\Omega_{ik}(b)$ corresponds to one-pomeron-exchange between states ϕ_i and ϕ_k written in the b-representation. So the low-mass diffractive dissociation cross section is

$$\sigma^{\text{D}}_{\text{low}M} \;=\; \sigma_{\text{el+SD+DD}} - \sigma_{\text{el}}, \tag{14}$$

where $\sigma_{\text{el+SD+DD}}$ corresponds to all possible low-mass dissociation caused by the 'dispersion' of the Good-Walker eigenstate scattering amplitudes.

Also, the pp elastic differential cross section of (8) becomes

$$\frac{d\sigma_{\text{el}}}{dt} \;=\; \frac{1}{4\pi}\left| \int d^2b \; e^{i\vec{q_t}\cdot\vec{b}} \sum_{i,k} |a_i|^2|a_k|^2 \; (1 - e^{-\Omega_{ik}(b)/2}) \right|^2, \tag{15}$$

where $-t = q_t^2$. As mentioned above, in order to correctly describe the dip region we must include the real part of the amplitude. We use a dispersion relation. For the even-signature pomeron-exchange amplitude this means

$$A \;\propto\; s^\alpha + (-s)^\alpha \qquad \text{and so we have} \qquad \frac{\text{Re}\;A}{\text{Im}\;A} = \tan(\pi(\alpha - 1)/2). \tag{16}$$

Formula (16) is transformed into b-space, so that the complex opacities can be constructed. For each value of b we calculate α and determine Re A from (16).

The coupling g_i of the pomeron to the ϕ_i eigenstates may be written

$$g_i = \gamma_i \sqrt{\sigma_0} F_i(t), \qquad (17)$$

where the form factors satisfy $F_i(0) = 1$, and $(\gamma_1 + \gamma_2)/2 = 1$. The cross section for the interaction of eigenstates ϕ_i and ϕ_k, via one-pomeron-exchange, is $\sigma_{ik} = \sigma_0 \gamma_i \gamma_k (s/s_0)^{\alpha_P - 1}$. The parameters of the form factors, $F_i(t)$, are adjusted to give, among other things, the observed behaviour of $d\sigma_{\rm el}/dt$ versus t, as shown in Fig. 1(b).

6. High-mass dissociation

In the absence of absorptive corrections, the $pp \to p + X$ cross section for dissociation into a system X of high mass M is

$$\frac{M^2 d\sigma_{\rm SD}}{dt dM^2} = \frac{g_{3P}(t) g_N(0) g_N^2(t)}{16\pi^2} \left(\frac{s}{M^2}\right)^{2\alpha(t)-2} \left(\frac{M^2}{s_0}\right)^{\alpha(0)-1}, \qquad (18)$$

where $g_N(t)$ is the coupling of the pomeron to the proton and $g_{3P}(t)$ is the triple-pomeron coupling. The value of the coupling g_{3P} is obtained from a triple-Regge analysis of lower energy data. However, g_{3P} is actually an effective vertex with coupling

$$g_{\rm eff} \;=\; g_{3P}\,\langle S^2 \rangle \qquad (19)$$

which already includes the suppression or survival factor $S^2(b) = \exp(-\Omega(b))$ – the probability that no other secondaries, simultaneously produced in the same pp interaction, populate the rapidity gap region, see (7). A more precise analysis [34] accounts for the survival effect $S^2_{\rm eik}$ caused by the eikonal rescattering of the fast 'beam' and 'target' partons. In this way, a coupling g_{3P} about a factor of 3 larger than $g_{\rm eff}$ is obtained, namely

$$g_{3P} \equiv \lambda g_N \qquad \text{with} \qquad \lambda \simeq 0.2, \qquad (20)$$

where g_N is the coupling of the pomeron to the proton.

7. Experimental evidence of the $k_t(s)$ effect

Can a simultaneous 'global' description be obtained, of all the high-energy elastic and diffractive data, accounting for all the puzzles listed in Section 3? This was the objective of [13], which was based on earlier observations in [7, 35]. This analysis was reviewed in detail in [14]. Here we sketch the main results.

Loosely speaking, one of the main puzzles, in the data sets, is that the diffractive dissociation cross sections observed at the LHC are smaller than expected from conventional RFT extrapolations of lower energy collider data. There is a good physics reason why this is so – the $k_t(s)$ effect [35, 13, 14].

7.1. Energy dependence of low-mass dissociation

We discuss, first, item (b) on the list; that is, the energy dependence of $\sigma_{\rm D}^{{\rm low}M_X}$. The pomeron-$|\phi_i\rangle$ coupling, g_i, is driven by the impact parameter separation, $\langle r \rangle$, between the partons in the $|\phi_i\rangle$ state. The well known example is so-called colour transparency, where the cross section $\sigma \propto \alpha_s^2 \langle r^2 \rangle$. However, if the transverse size of the pomeron, ($\sim 1/k_t$), becomes much smaller than this separation, then the cross section (and coupling) will be controlled by the pomeron size, that is by the characteristic k_t in the pomeron ladder, which we denote by k_P. In this limit $\sigma \propto 1/k_P^2$. Therefore it is natural to choose the following parametrization for the γ_i in the pomeron-$|\phi_i\rangle$ couplings of (17)

$$\gamma_i \propto \frac{1}{k_P^2 + k_i^2}, \tag{21}$$

where k_P^2 is expected to have an energy dependence

$$k_P^2(s) = k_{P0}^2 \left(\frac{s x_0^2}{s_0}\right)^D . \tag{22}$$

In other words, during the evolution in $\ln(1/x)$, due to the diffusion in $\ln k_t^2$, k_P^2 grows approximately as a power D of $1/x$. Of course, we do not expect that the whole available $\ln(1/x)$ (rapidity) space will be subject to diffusion. Rather, we assume, that as x decreases, the diffusion starts from some relatively low $x = x_0$ parton with $x_0 = 0.1$. That is, the rapidity space available for the $\ln k_t^2$ diffusion is not $\ln(s/s_0)$, but is diminished by $\ln(1/x_0)$ from both sides. (As usual we use $s_0 = 1$ GeV^2.) The typical transverse momentum of this (starting) parton, inside the state ϕ_i, is denoted by the parameter k_i in (21).

The parametrisation of γ_i in (21) is such that at very large energies all the $\gamma_i \to 1$. That is, the interaction will not destroy the coherence of the wave functions of the incoming protons. In other words, $\sigma_{\rm D}^{{\rm low}\,M_X}$ will decrease with energy, while at lower energies we tend to the naive expectation $\gamma_i \propto 1/k_i^2$. In the 'global' description [13], the dissociation is slowed sufficiently with increasing energy such that the values of the cross section $\sigma_{\rm D}^{{\rm low}\,M_X}$ are compatible with the data – namely, the model gives 2.6 mb at $\sqrt{s} = 62.5$ GeV, and 3.8 mb at $\sqrt{s} = 7$ TeV. The latter value is consistent with the TOTEM measurement of 2.6±2.2 mb.

7.2. Energy dependence of high-mass dissociation

An analogous puzzle was listed in item (c) of Section 3, so it is natural to introduce some energy dependence into the triple pomeron coupling λ, of (20), which drives high-mass dissociation. As before, it arises from the characteristic transverse momenta of the intermediate partons inside the pomeron ladder (i.e. the size of the pomeron), which depend on the rapidity (or energy) of corresponding partons, It looks natural to take

$$\lambda \propto 1/k_P^2(s) , \tag{23}$$

where $k_P^2 = \langle k_t^2 \rangle$. The diffusion in $\ln k_t^2$ occurs from both the beam and target sides of the ladder. Following (22), we take $k_P^2 \propto (x_0/x)^D$ for diffusion from one side and $k_P^2 \propto (x_0/x')^D$ from the other side; where to evaluate x' we use the relation $xx's = \langle m_T^2 \rangle$, and assume $\langle m_T^2 \rangle = s_0 = 1$ GeV^2. So we parametrize $k_P(s)$ by

$$k_P^2(s) = k_0^2 \left(\left(\frac{x_0}{x}\right)^D + \left(\frac{x_0}{x'}\right)^D \right) , \tag{24}$$

where we take the same parameter D and evolve from the same starting point $x_0 = 0.1$ as (22) for γ_i of (21). We calculate x' as $x' = s_0/xs$ with $s_0 = 1$ GeV^2. If $x > x_0$ then x_0/x ratio is replaced by 1, and similarly for x'. The global description [13] gives $D = 0.28$.

With this energy dependence of the coupling λ, the values obtained in [13, 14] for the single proton dissociation cross section (integrated over the three mass intervals used by TOTEM [19]) are shown in Table 1. We see that the agreement with the mass dependence of the TOTEM data is now satisfactory. Moreover, note that $\lambda = 0.18$ at relatively low energies, when both $x > x_0$ and $x' > x_0$, such that λ ceases to be energy dependent. This is in agreement with the value 0.2 of (20) obtained in the triple-Regge analysis.

Mass interval (GeV)	(3.4, 8)	(8, 350)	(350, 1100)
Prelim. TOTEM data	1.8	3.3	1.4
CMS data		4.3	
Model [13]	2.3	4.0	1.4

Table 1. The values of the cross section (in mb) for single proton dissociation (integrated over the three indicated mass intervals) as observed by TOTEM [19], compared with the values obtained in the model of [13, 14]. Recall that TOTEM claims that their preliminary measured cross sections have a 20% uncertainty. Note that the value quoted for the CMS [21] cross section of dissociation is integrated over the 12 - 394 GeV M_X interval (close to, but in terms of $\ln M_X$, a bit smaller than, the interval (8 - 350) GeV chosen by TOTEM).

7.3. An interesting and unusual effect on the energy behaviour of $\sigma_{\rm tot}$

Allowing the pomeron couplings to be energy dependent produces the unusual energy behaviour of the total cross section noted in item (a) of Section 3. Indeed, the observation that the $\gamma_i \to 1$ at high energy, means that the growth of the total cross section speeds up in the CERN-ISR $\to$ LHC $\to$ 100 TeV interval. For example, if we followed the conventional 'classic' Regge approach and assumed that the couplings γ_i were independent of energy, and fix the values corresponding to the CERN-ISR energy of $\sqrt{s}$ = 50 GeV, then we obtain $\sigma_{\rm tot}$ = 74, 91 and 134 mb for $\sqrt{s}$ = 1.8, 7 and 100 TeV respectively. On the other hand, allowing the couplings to be energy dependent, as follows from (22), the values of $\sigma_{\rm tot}$ grow faster with energy, namely. 77, 99 and 166 mb, and now in accord with the LHC measurement.

These numbers demonstrate an important new fact. That is, that the energy dependence of the total and elastic cross sections is not only driven by the parameters of the pomeron trajectory, but also by the energy behaviour of the factors γ_i; in other words by the decomposition of the proton-pomeron coupling between the different Good-Walker eigenstates. Note that this 'acceleration' of the total cross section growth, due to the variation $\gamma_i \to 1$, takes place in only one energy interval. We are fortunate to observe it in just the Sp$\bar{p}$S – LHC collider interval.

7.4. Factorisation for the diffractive cross sections?

Concerning item (d), we note that TOTEM have measured $\sigma_{\rm el}$, $\sigma_{\rm SD}$ and $\sigma_{\rm DD}$ in some kinematic interval. If we take the simplest pomeron exchange diagrams that describe these processes, then we obtain the naive factorisation relation $\sigma_{\rm DD}\sigma_{\rm el}/(\sigma_{\rm SD})^2 = 1$, whereas the data give $0.116 \times 25/(0.9)^2 \simeq 3.6$, integrated over the same kinematic region. Here $\sigma_{\rm SD}$ is the single dissociation cross section from *one* proton, not the sum of both dissociations. This discrepancy is not a surprise. We expect violation of the naive result due to the different t-slopes and survival factors for the processes. These, together with a more careful treatment of the calculation of $\sigma_{\rm DD}$, removes the discrepancy [13].

8. Tension between the high-mass single dissociation data

Although TOTEM have made the most detailed observations of high-mass single proton dissociation in high energy pp collisions, the 'global' diffractive model [13] was tuned to simultaneously describe the TOTEM data *together* with earlier measurements of single dissociation. Formally the diffractive dissociation data from different groups do not contradict each other, since they are measured for different conditions. However, global fits of all diffractive data reveal the tensions between the data sets [13, 14, 37]. Indeed, it is not easy to reconcile the preliminary TOTEM measurements for $\sigma_{\rm SD}$ with (a) the 'CDF' measurement of single dissociation and (b) the yield of Large Rapidity Gap events observed by ATLAS and CMS.

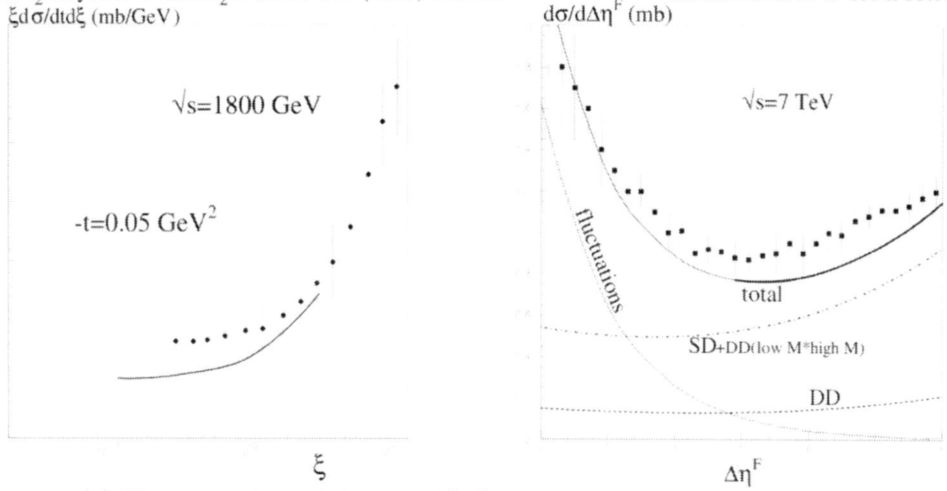

Figure 2. (a) The comparison of the model [13] with data for single proton dissociation measured by the CDF collaboration, given in [25] but not including a normalisation uncertainty of about 10-15%. The inclusion of the secondary Reggeon contribution RRP is responsible for the rise of the curve as ξ increases. (b) The ATLAS [20] measurements of the inelastic cross section differential in rapidity gap size $\Delta\eta^F$ for particles with $p_T > 200$ MeV. Events with small gap size ($\Delta\eta^F \lesssim 5$) may have a non-diffractive component which arises from fluctuations in the hadronization process [36]. This component increases as $\Delta\eta^F$ decreases (or if a larger p_T cut is used [36, 20]). The data with $\Delta\eta^F \gtrsim 5$ are dominantly of diffractive origin, and are compared with the 'global' diffractive model [13]. The DD contribution of events where both protons dissociate, but the secondaries produced by one proton go into the beam pipe and are not seen in the calorimeter, is shown by the dashed curve. The figures are taken from [13].

Basically the global description of [13] overestimates the TOTEM data and underestimates the CDF [25], ATLAS [20] and CMS [21] data. This can be seen in Fig. 2 and from Table 1.

9. Discussion

We have discussed a model [13] in which all the high-energy diffractive data may simultaneously be described within the Regge Field Theoretic framework based on only one pomeron pole. However, to reach agreement with the data, the model includes pomeron-pomeron interactions, arising from multi-pomeron vertices, and allows for the $k_t(s)$ dependence of the multi-pomeron vertices. Recall that, due to the BFKL-type diffusion in $\ln k_t^2$ space, together with the stronger absorption of low k_t partons, the typical transverse momentum, k_t, increases with energy depending on the rapidity position of the intermediate parton or the multi-pomeron vertex. This $k_t(s)$ effect enables the model to achieve a relatively low probability of low-mass dissociation of an incoming proton and to reduce the cross section of high-mass dissociation in the central rapidity region in comparison with that observed closer to the edge of available rapidity space – both of which are features demanded by the recent TOTEM data.

The model [13] can be refined when more precise and consistent diffractive data become available. At present the parameters are *tuned* to give a reasonable description of all aspects of the available diffractive data. If, instead, a χ^2-fit to the data had been performed, then the few dissociation measurements of TOTEM (the preliminary values of $\sigma_{\rm SD}$ with 20% errors in three mass intervals, and one value of $\sigma_{\rm DD}$) would have carried little weight. On the other hand, all the TOTEM data are self-consistent between themselves. Moreover, these data reveal a very reasonable tendency of the $d\sigma_{\rm SD}/d\xi$ dependence, close to that predicted in [7] where the k_t distribution of the intermediate partons inside the pomeron ladder, and the role of the transverse size of the different QCD pomeron components, were accounted for more precisely. Therefore, the model [13] was tuned to achieve a 'compromised' description of the data sets in 'tension'. On the theory side, the model could be improved by including a more complicated (non-local) structure of the original pomeron. At present, the model considers just an 'effective' pomeron pole renormalized by enhanced absorptive corrections.[1]

Acknowledgements
I thank Olga Piskunova for arranging a very fruitful and enjoyable Workshop, and Misha Ryskin and Valery Khoze for many enjoyable collaborations and valuable discussions on the subject of the article.

References
[1] Kaidalov A B 1971 *Sov.J.Nucl.Phys.* **13** 401
[2] Kaidalov A B, Khoze V A, Pirogov Y F and Ter-Isaakyan N L 1973 *Phys.Lett.* **B45** 493–496
[3] Kaidalov A B and Ter-Martirosian K A 1982 *Phys.Lett.* **B117** 247–251
[4] Kaidalov A B, Ter-Martirosian K A and Shabelski Y 1986 *Sov.J.Nucl.Phys.* **43** 822–826
[5] Kaidalov A B, Ponomarev L A and Ter-Martirosian K A 1986 *Yad.Fiz.* **44** 722–728
[6] Ryskin M G, Martin A D and Khoze V A 2009 *Eur.Phys.J.* **C60** 249–264 (*Preprint* 0812.2407)
[7] Ryskin M G, Martin A D and Khoze V A 2011 *Eur.Phys.J.* **C71** 1617 (*Preprint* 1102.2844)
[8] Maor U 2013 (*Preprint* 1310.7340)
[9] Gotsman E 2013 (*Preprint* 1304.7627)
[10] Ostapchenko S 2010 *Phys.Rev.* **D81** 114028 (*Preprint* 1003.0196)
[11] Kaidalov A B and Poghosyan M G 2011 (*Preprint* 1109.3697)
[12] Gribov V N 1968 *Sov.Phys.JETP* **26** 414–422
[13] Khoze V A, Martin A D and Ryskin M G 2013 (*Preprint* 1312.3851)
[14] Khoze V A, Martin A D and Ryskin M G 2014 (*Preprint* 1402.2778)
[15] Antchev G *et al.* (TOTEM) 2011 *Europhys.Lett.* **95** 41001 (*Preprint* 1110.1385)
[16] Antchev G *et al.* (TOTEM) 2013 *Europhys.Lett.* **101** 21002
[17] Antchev G *et al.* (TOTEM) 2013 *Europhys.Lett.* **101** 21003
[18] Antchev G *et al.* (TOTEM) 2013 *Phys.Rev.Lett.* **111** 262001 (*Preprint* 1308.6722)
[19] Oljemark F and Osterberg K (TOTEM) *Studies of soft single diffraction with TOTEM at $\sqrt{s}=7$ TeV; LHC students poster session, 13 March 2013*
[20] Aad G *et al.* (ATLAS) 2012 *Eur.Phys.J.* **C72** 1926 (*Preprint* 1201.2808)
[21] CMS 2013 Measurement of diffraction dissociation cross sections at $\sqrt{s}$=7 TeV at the LHC, CMS-PAS-FSQ-12-005
[22] Abelev B *et al.* (ALICE) 2013 *Eur.Phys.J.* **C73** 2456 (*Preprint* 1208.4968)
[23] Amos N A *et al.* (E710) 1990 *Phys.Lett.* **B247** 127–130
[24] Abe F *et al.* (CDF) 1994 *Phys.Rev.* **D50** 5518–5534
[25] Goulianos K A and Montanha J 1999 *Phys.Rev.* **D59** 114017 (*Preprint* hep-ph/9805496)
[26] Schegelsky V A and Ryskin M G 2012 *Phys.Rev.* **D85** 094024 (*Preprint* 1112.3243)
[27] Donnachie A and Landshoff P V 1992 *Phys.Lett.* **B296** 227–232 (*Preprint* hep-ph/9209205)
[28] Bozzo M *et al.* (UA4) 1984 *Phys.Lett.* **B147** 385
[29] Augier C *et al.* (UA4/2) 1993 *Phys.Lett.* **B316** 448–454
[30] Arnison G *et al.* (UA1) 1983 *Phys.Lett.* **B128** 336
[31] Antchev G, Aspell P, Atanassov I, Avati V, Baechler J *et al.* 2011 *Europhys.Lett.* **96** 21002 (*Preprint* 1110.1395)
[32] Amaldi U, Jacob M and Matthiae G 1976 *Ann.Rev.Nucl.Part.Sci.* **26** 385–456
[33] Good M L and Walker W D 1960 *Phys.Rev.* **120** 1857–1860
[34] Luna E G S, Khoze V A, Martin A D and Ryskin M G 2009 *Eur.Phys.J.* **C59** 1–12 (*Preprint* 0807.4115)
[35] Khoze V A, Martin A D and Ryskin M G 2013 *Eur.Phys.J.* **C73** 2503 (*Preprint* 1306.2149)
[36] Khoze V A, Krauss F, Martin A D, Ryskin M G and Zapp K C 2010 *Eur.Phys.J.* **C69** 85–93 (*Preprint* 1005.4839)
[37] Ostapchenko S 2014 (*Preprint* 1402.5084)

[1] The model (which includes eikonalized absorption, but not enhanced absorption) has an effective pole with a trajectory $\alpha_P(t) \simeq 1.12 + 0.05t$, intermediate between the effective soft pomeron with $\alpha_P(t) \simeq 1.08 + 0.25t$ of (1) and the 'bare' QCD pomeron with $\alpha_P(t) \simeq 1.25 + 0t$, in GeV units.

On dynamics of hard elastic scattering of hadrons

O.V. Kancheli

Institute of Theoretical and Experimental Physics,
B. Cheremushinskaya 25, 117259 Moscow, Russia

E-mail: kancheli@itep.ru

Abstract. The main contribution to hard elastic scattering at high energies comes from components of wave functions of colliding hadrons that contain minimal number of partons. We discuss the details of such a mechanism in the regge and parton approaches and estimate the probability that colliding hadrons are in bare states containing only valent partons. The behavior of cross sections in this regime at various energies can give nontrivial information on high energy dynamics.

1. Introduction.

The elastic scattering of hadrons at high energies is reasonably well described in the regge approach. At small t it has the diffractive nature and is connected with the pomeron exchanges (fig.1). As confirmed by both experimental data and QCD calculations the pomeron $\mathbb{P}$ is supercritical, i.e. its intercept $\alpha_P = 1 + \Delta$ is higher than unity, $\Delta > 0$. For the supercritical pomeron the mean number of $\mathbb{P}$ exchanges increases with energy as $\sim s^{\Delta}$. When energy increases it results in gradual change of energy behavior of total cross section from the power growth, $\sigma_{tot} \sim \exp(\Delta y)$ to the Froissart type behavior, $\sigma_{tot} \sim y^2/m^2$, where $y = \log s/m^2$ is full rapidity.

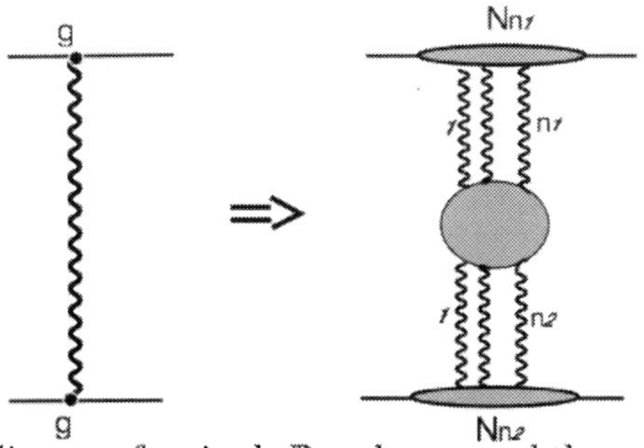

Fig.1 : The reggeon diagram for single $\mathbb{P}$ exchange and the general reggeon diagram for elastic amplitude, which becomes important at very high energy.

The width of the diffraction peak due to the single $\mathbb{P}$ pole exchange, $\delta t \sim 1/R^2(y)$, is determined by the regge radii of colliding hadrons, $R^2(y) \simeq 2R_0^2 + \alpha' y$, and the elastic amplitude decreases exponentially with $k_\perp^2 = -t$ as $\sim \exp[-k_\perp^2 R^2(y)]$. At asymptotical energies where the Froissart regime starts the interaction radius grows logarithmically, $R(y) \sim y$, due to interplay of alternating regge cut contributions.

The parton content of the pomeron can be realized as a gluonic ladder (fig.2) where gluons are uniformly distributed in rapidity. The mean rapidity gap between gluon steps, $\delta_y \sim 1/\alpha_s(\langle k_{i\perp}^2 \rangle)$, is determined by the value of QCD coupling constants on the scale of average transverse momenta of intermediate gluons $\langle k_{i\perp}^2 \rangle \simeq -t$.

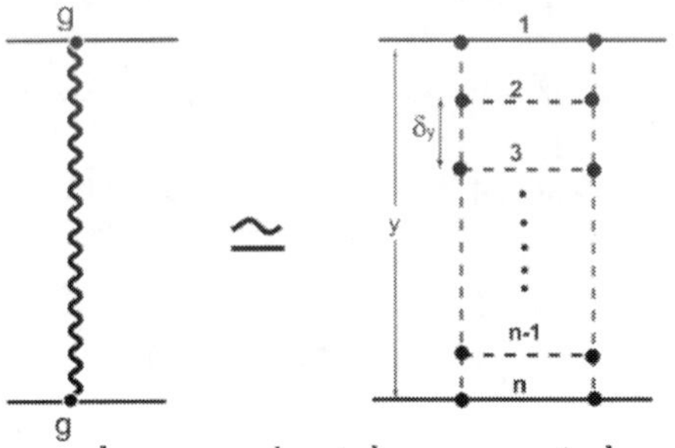

Fig.2 : The pomeron can be approximately represented as the BFKL type gluon ladder with the average rapidity interval between steps δ_y, so that the average $n \simeq y/\delta_y$.

The mean number of intermediate gluons in the pomeron is $\simeq y/\delta_y$. The distribution in number of gluons is the Poissonian one, so the probability to have no intermediate gluons is exponentially small, $\sim \exp[-y/\delta_y]$. With increase of $k_\perp$ the character of elastic scattering gradually changes from diffractive mechanism to the perturbative parton exchange.[1] The trajectory $\alpha_P(t)$ of the QCD pomeron at $t = 0$ is higher than unity. It is nonlinear and at large $-t = k_\perp^2$, where the perturbative mechanisms dominate, must behave in perturbative way as $\alpha_P(k_\perp) \simeq 1 + c/\log(k_\perp^2)$.

In terms of the regge approach when $|t|$ increases at fixed energy the cut contributions with larger and larger number of pomeron exchanges dominate and the elastic amplitude can be approximately represented by the Orear type expression $\sim \exp[-k_\perp f(y)]$. In this case the large total transferred momentum $k_\perp$ is distributed more or less evenly among the exchanged pomerons, and this is connected with the approximate linearity of regge trajectories at low t values.

For larger $k_\perp$ this uniform distribution of transferred momentum between pomeron lines changes, and the large transferred momentum $k_\perp$ is concentrated mainly on a single pomeron line. This is primarily related with the nonlinearity of the pomeron trajectory and its hard

[1] Note that, as can be seen from a comparison with the experimental data, the regge trajectories look linear up to rather large $k_\perp^2 \simeq 2 \div 5 GeV^2$, and only after that they can move to the perturbative values. The details of this phenomenon are not so well investigated, and it is, probably, related to the large nonperturbative contributions, essential up to sufficiently small distances $\sim 0.1 \div 0.2\ fm$.

The structure of pomeron in QCD can be rather complicated and $\mathbb{P}$ probably consists of a sequence of regge poles $\mathbb{P}_n$ with intercepts $\simeq 1 + c/n$. They have small t slope $\sim 1/n^2$ and their internal virtualities $< q_\perp^2 >_n$ grow like $\sim \exp n$. At high $-t$ all these satellite $\mathbb{P}_n$ accumulate at their bare values like $\alpha_{nP}(k_\perp) \simeq 1 + c/(n + c_1 \log(k_\perp^2/\Lambda_c^2))$. At high momentum transfer the contribution of satellites $\mathbb{P}_n$ with high n in the amplitude can be even dominant – so that the essential $n \sim \log(-t)$. But here we do not take into account these details because all $\mathbb{P}_n$ trajectories move at high $-t$ to the same bare value, and the structure of exchange is determined by the minimal perturbative graph.

satellites at large $k_\perp$.

As a result, the t-behavior of elastic scattering amplitude is determined by the pomeron vertices, and one can expect the power behavior $A \sim 1/k_\perp^{2\nu}$ of the full amplitude, where ν depends on the minimal number of exchanges needed to scatter all parton components of hadron on the same angle.

In this mechanism (when the $\mathbb{P}$ trajectory is already frozen) the main contribution to the hard scattering amplitude comes from components of wave functions of the colliding hadrons with minimal possible number of partons [2]. This high $k_\perp$ scattering mechanism contains two main ingredients:

- The first is the amplitude of hard scattering of valent (bare) constituents. It is determined by the perturbative QCD, and can be estimated even on the dimensional grounds.
- The other is a probability to find the hadron in such a "bare" state with the minimal number of partons, and it is determined mainly by nonperturbative physics.

What one needs first of all is to know the dependence of this probability on the hadron energy and its transverse resolution. This is what we consider below in the regge and parton approaches.

2. Cross sections

Consider the contribution to the elastic scattering amplitude $A(s,t)$ from rather general reggeon diagrams of Fig.3 at high $|t| \gg 1/R^2$, but such that $|t| \ll s$. Because (as we have assumed) the pomeron trajectory is nonlinear, and freezes at high $-t$, the $\mathbb{P}$ exchange amplitude $v \sim ig^2(t)s^{\alpha(t)}$ behaves at high $-t$ in power-like manner. As a result in this case the main contribution to A

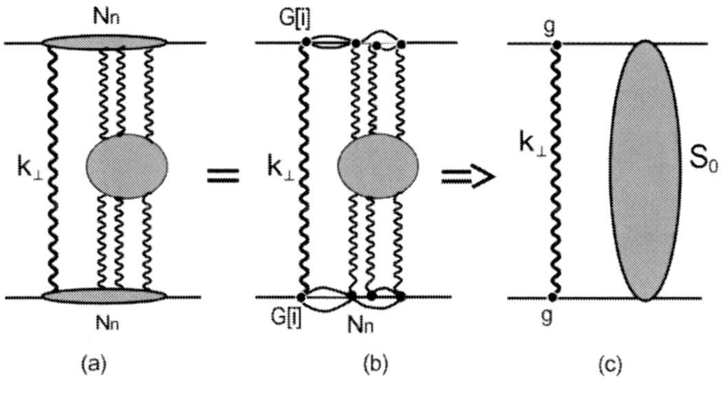

Fig.3 : **(a)** General reggeon diagram for elastic multi-$\mathbb{P}$ amplitude, where the $\mathbb{P}$ line with high transferred momentum is separated off.
(b) The decomposition of the multipomeron vertices N_n into jets.
(c) The diagram essential at large $k_\perp$.

comes from diagrams of Fig. 3a in which almost whole transverse momentum is transferred trough a single $\mathbb{P}$ line. The full diagram for A can be symbolically represented as in Fig.3c

[2] The picture closely corresponds to the quark counting rules [1] and its development [2]-[3]. The power behavior of elastic cross-sections at large -t is approximately seen in experiments.

which contribution is

$$A(y,t=-k_\perp^2) = \int d^2q_{1\perp} S(y,q_\perp^2) v(y,(k+q)_\perp^2) \simeq$$
$$\simeq v(y,k_\perp^2) \int d^2q_{1\perp} S(y,q_\perp^2) = v(y,k_\perp^2)\widetilde{S}(y,0) , \qquad y = \log s/m^2 , \tag{1}$$

where we separated off the integration over the $\mathbb{P}$ line with high momentum transfer. Here $\widetilde{S}(y,b)$ is the elastic S-matrix in impact parameter plane b. It appears in (1) at zero impact parameter, reflecting dominance of small values of b at high momentum transfer. This finally gives

$$\frac{d\sigma}{dt} \simeq S_0^2(y)\,\frac{d\hat{\sigma}}{dt} , \qquad S_0(y) \equiv |\widetilde{S}(y,b=0)| , \tag{2}$$

where $d\hat{\sigma}(y,t)/dt$ is the "bare" cross-section, connected with hard exchange $v(y,k_\perp^2)$. The quantity $\widetilde{S}(y,b=0)$, entering (1), represents the contribution from the interaction in the initial and final state, and the $\widetilde{S}(s,b)$ is equal to the amplitude that one of colliding particles moves trough another without interaction at given s and impact parameter b. One factor $S_0(y)$ in (2) can be interpreted as a probability to find the fast particles in such a "bare" state that they almost do not interact with a target. Another factor $S_0(y)$ reflects that scattered particles in the final state are also in such a bare state. In the lab frame of one of colliding particles the $S_0(y)$ also gives the probability that another particle is in the state without a soft parton cloud.

In the factorized form (2) for $d\sigma/dt$ the quantity S_0^2 represents the contribution of large transverse distances and $d\hat{\sigma}/dk_\perp^2$ – of the small ones.

The essential point which has to be taken into account, when going from diagrams Fig.3a,b to Fig.3c, is connected with the structure of multipomeron vertices N_n, when some pomeron lines in N_n have large transverse momenta. In general, all intermediate diffractive states (the states corresponding to cutting the diagram between pomerons) can contain either a single particle (elastic rescatterings which correspond to the eikonal approximation) or several particles (nonelastic diffractive rescatterings). So in resulting expressions (1), (2) one should sum over the states [i] of such jets. But the particles produced in pomeron vertices $G[i]$ due to hard diffraction would have transverse momenta of order $k_\perp$ much larger than average ones. The successive diffractive rescatterings of these high-$k_\perp$ states will lead to an additional smallness $O(1/k_\perp^2)$ of corresponding contribution to the amplitude. Alternatively, this jet can be especially aligned in order to compensate the large transferred momentum in next rescatterings – the corresponding small probability results in the same small factor in multiparticle jet contribution. Therefore at high $k_\perp$ in the states outgoing from the hard vertices g in Fig.1c only one-particle state survives. Such an answer looks natural for the reason that in this case the minimal number of constituents pass trough the hard vertex.

The cross-section for the hard $\mathbb{P}$ -exchange $d\hat{\sigma}(y,t)/dt \sim g_1^2(t)g_2^2(t)$ is mainly determined by the behavior of $\mathbb{P}$ -vertices $g_j(t)$, because at large $-t$ the purely reggeon part

$$s^{\alpha(t)-1} \sim \exp(\ c\ y\Delta_P/\log(-t)\ ) , \qquad c \sim 1 \tag{3}$$

grows very slowly with s. In perturbative QCD the hard part of the pomeron can be approximately represented as the BFKL ladder with mean interval between rapidities of emitted gluons of order of $\sim 1/\alpha_s(t)$. Such a BFKL type $\mathbb{P}$ exchange has almost the same form as the direct 2n gluon exchange; the $n > 1$ contributions can arise because the BFKL pomeron contains also the multigluon t-channel contributions due to gluon reggeization [3]. On this way we come finally to the quark counting type model.

[3] At high $-t$ not only the pomeron but also the other multigluon QCD reggeons can equally contribute to the

The behavior of $g_i^2(t)$ at large $|t|$ in QCD can be estimated even quasi-classically, and this approach leads finally to the t dependence:

$$d\hat{\sigma}(y,k_\perp)/dk_\perp^2 \sim (\alpha_s(k_\perp))^\nu/(k_\perp^2)^N \quad , \tag{4}$$

with

$$\nu = n_1 + n_2 + |n_1 - n_2| \quad , \qquad N = \frac{1}{2}(3(n_1+n_2) + |n_1 - n_2|) - 1 \quad , \tag{5}$$

where numbers N and ν depend on the number of valent constituents n_1 and n_2 in colliding hadrons (in fact the fast quarks), and on a configuration in which the hard scattering of constituents takes place. In (4) we have neglected the regge factor (3), which at high $-t$ grows very slowly with s, and have also neglected the possible Sudakov type suppression factors – they can even approximately compensate one another.

The simplest configuration is such that before scattering all constituents in both hadrons are gathered in the small transverse region of size $\sim 1/k_\perp$ and then these bunches scatter on one another at mean transverse distance $\sim 1/k_\perp$. In this case we have

$$\nu = n_1 + n_2 + |n_1 - n_2| \quad , \qquad N = 2n_1 + 2n_2 - 2 \quad , \tag{6}$$

and we have from (4) the behavior of $\sim 1/t^{10}$ type for pp and to $\sim 1/t^8$ for πp cases.

There is also another initial transverse configuration of constituents, in which the scattering index N entering Eq.(4) is less then in (6). It corresponds to a scattering in a state in which the hadron constituents are arranged (with an accuracy $\sim 1/k_\perp$) in the transverse plane on the line perpendicular to a scattering plane [4]. In such a case we have

$$N \;\to\; \hat{N} \;=\; \frac{5}{4}\,|n_1+n_2| + \frac{3}{4}\,|n_1 - n_2| - \frac{1}{2} \quad , \tag{7}$$

which leads to the $\sim 1/|t|^7$ behavior for pp and to $\sim 1/|t|^{13/2}$ for πp cases.

Note that the bare hadron cross section $d\hat{\sigma}/dk_\perp^2$ given by (4) can be approximately interpreted as the cross section of specific quasi-elastic process, where in the final state we have two hadrons with high transverse momenta $\simeq \pm k_\perp$, and all other particles have small $k_{i\perp}^2 \sim \langle k_\perp^2 \rangle$. The particles which have large transverse momenta $k_\perp^2 \gg \langle k_\perp^2 \rangle$ and take more than a half of the total energy originate from hard interactions of leading partons. All other final particles are soft and appear in configurations typical for mean inelastic events at the same energy. These particles are created by the standard hadronization of soft parton cloud, which follows after "removing" the leading partons in a hard interaction.

Such events also resemble the final state containing two high energy jets in special configurations, in which all the jet energy is concentrated on one fast particle (pion or nucleon). The probability of jet coming in such "empty" state very likely contains the same damping factors $d\hat{\sigma}/dk_\perp^2$ and $S_0(s)$.

hard exchange in the same way as the BFKL pomeron $\mathbb{P}$ and odderon. Probably all such n-gluon trajectories $\alpha_n(t)$ at large $-t$ are approximately degenerated and accumulate near intercept equal one. The difference in their contributions comes only from different values of corresponding regge vertices.

[4] In fact all what is needed and what leads to expression (7) is the following: all valent constituents must scatter at very close angles $\theta \simeq k_\perp/k_z$ such that all relative transverse momenta of scattered partons be $\lesssim m$. Also one must fulfil the condition that after scattering all the constituent partons should be located in the same packet of longitudinal size $\sim 1/k_z$. It leads to the condition that hadrons predominately scatter in specific initial configurations, when their partons are arranged on the lines in transverse planes. These lines must be perpendicular to the scattering plane, but the relative separation of partons on these lines can be arbitrary. Such a picture of hard elastic scattering picture looks in a sense similar to the Landshoff mechanism [3].

It can be interesting to detect similar events in high-energy heavy ion interactions. For example, in A_1A_2 nucleus collisions those are the events with two nucleons of high transverse momenta $\pm k_\perp$ (but without jets) and all other hadrons with transverse momenta typical for soft A_1A_2 collisions. The cross-section of such a process is

$$d\sigma/dk_\perp^2 \simeq A_1A_2 \cdot d\hat{\sigma}/dk_\perp^2 \ ,$$

because the nuclei are transparent for the bare nucleons.

3. The estimation of $|S_0|^2$ in parton models

Thus, to understand the large $-t$ behavior of $d\sigma/dt$ it is essential to estimate the behavior of $|S_0|$, which gives the probability for fast hadron to be in a bare state, i.e. to contain only valent components and no additional parton cloud. One can expect that this probability is small because at all accessible energies the fast hadron is basically accompanied by the soft parton cloud related to $\mathbb{P}_n$ exchange.

The $\mathbb{P}$-ladder corresponds, roughly speaking, to a soft parton (gluon) cascading with some mean step in rapidity δ_y. Because the cascading steps are almost independent, the probability to have no cascade (i.e. to realize a state without additional partons) is of Poisson type $\sim \exp(-y\widetilde{\Delta})$, where $\widetilde{\Delta} = c/\delta_y$, $c \sim 1$. It corresponds to the fact that at the rapidity boost the mean number of low energy partons $n(y)$ is defined by the linear equation $\partial n(y)/\partial y = \widetilde{\Delta}\, n(y)$ [5]. Thus the crucial quantity is $\widetilde{\Delta} \simeq \delta_y^{-1}$ where δ_y is the mean rapidity step between essential degrees of freedom in the pomeron ladder. This gives an estimate $S_0(y) \simeq \exp(-y/\delta_y)$.

These arguments can be presented in a slightly different way. To scatter at large $k_\perp$ both hadrons should have the minimal parton clouds. It means that colliding particles must fluctuate to a state with small transverse size $r(y)$. The probability w of such a fluctuation is $\sim (r(y))^{2(\nu-1)}$, where ν is a number of the valent constituents. In this case, the mean rapidity interval between steps in parton ladder would be $\delta_y \sim 1/\alpha_s(r(y))$. From the condition $\delta_y \simeq y$ (no partons except of valent ones in the whole rapidity interval) we find that $r(y) \sim \exp(-c_1 y)$, and we come to the same type of the exponential dependence $S_0(y) \sim w(y) \sim \exp(-2c_1(\nu-1)y)$ as before [6].

Additional arguments for exponential behavior of probability $S_0(y)$ come from the requirement of boost invariance of hard collision description in partonic terms. Let us consider this process in an arbitrary longitudinal frame, when the colliding particles have rapidities y_1 and y_2. Then the quantity $S_0(y)$, which is the amplitude that both colliding particles are in a bare state must have the multiplicative form

$$S_0(y) = S_0(y_1) \cdot S_0(y_2) \ ,$$

where $y = y_1 + y_2$. Because at boost $y_1 \to y_1 + \eta$, $y_1 \to y_1 - \eta, y \to y$, the only functional form of $S_0(y)$ that satisfies this condition in arbitrary frame is the exponential one $S_0(y) \sim \exp(-cy)$.

So to better estimate the behavior of $S_0(y)$ we need to know the mean number of steps in the pomeron ladder. Two possibilities can be emphasized, and they correspond to a different choice of partons – independent degrees of freedom in the Fock wave function of fast hadron.

[5] Evidently this is true only for supercritical pomeron. If $\alpha_P(0) < 1$ the probability to find the hadron in the "bare" state is finite and does not decrease with the hadron energy.

[6] This type of reasoning can remind the approach [6] to high energy scattering used in dual models embedded in the AdS_5 space.

In one case we can consider the pomeron ladder as constructed from colorless particles (for example, from π and ρ mesons), as it is usually done in various multiperipheral type models. In this case the steps (parton) density in rapidity is $\simeq 0.5\rho \simeq 1$, where ρ is the mean density of produced hadrons at not too high energies, when single pomeron exchange dominates.

In other case one can identify the soft pomeron ladder as a sparse gluon ladder with large steps $\widetilde{\delta}_y \simeq \Delta_P^{-1}$, where $\Delta_P \simeq 0.1 \div 0.2$ is the "experimental" soft pomeron intercept [7].

I guess the second possibility is theoretically more preferable, because the colorless hadron-partons do not represent the independent degrees of freedom, and therefore they are not completely appropriate to be used in the Fock space Hamiltonian.

If hard partons were also essential in the Fock wave function then at first sight instead of $\simeq \exp(-y/\delta_y)$ one can expect a more complicated behavior of $|S_0(y)|$. But the hard gluons are mainly taken into account in the parton w.f. as constituents of soft partons. We come to the same conclusion from a different way by remarking that the hard gluon spectra "measured" in deep-inelastic reactions can be successfully described [5] as coming from renormalization group rescaling of the soft parton component.

At asymptotic energies the saturation of soft and even more hard partons can become fully essential and as a result the saturation scale $Q^{(sat)}(y)$ in transverse momenta can also become large. Then one should take into account in $|S_0|^2$ also more hard partons. Note that even in the asymptotical Froissart regime, where the hard saturation probably dominates, the behavior of $|S_0|$ can be estimated in the same way – as the probability that the valent components do not emit any primary gluons in the corresponding ranges of energy and transverse momenta ($q_\perp < \langle k_\perp \rangle \sim Q^{(sat)}(y)$). This is quite enough, because all other partons are emitted by these primaries. This evidently leads to a mean number of the primary partons

$$\bar{n}(y, k_\perp) \sim \int d\omega/\omega \int^{k_\perp} dq_\perp^2 \alpha_s(q_\perp)/q_\perp^2 \ \sim \ y \ln\ln(k_\perp/m) \ ,$$

and to Sudakov-type factor for the no-emission probability $W(y, k_\perp) \sim \exp(-\bar{n})$, and, eventually, to the estimate $|S_0(y, k_\perp)| \simeq W(y_1, k_\perp)W(y_2, k_\perp)$, which is again boost-invariant, and has almost the same behavior.

4. The estimation of $|S_0|^2$ in regge approach

The numerical values of $S_0 \equiv |\widetilde{S}_0(y, b = 0)|$ can be estimated directly from the experimental data on the behavior of the profile function $F(y, b) = 1 - \widetilde{S}(y, b)$ at $b = 0$, calculated by the Fourier transformation of $\sqrt{(d\sigma/dt)_{(exp)}}$. For the pp and $p\bar{p}$ scattering it ranges from values $S_0 \simeq 0.6$ at $\sqrt{s} \sim 50 GeV$ up to values $S_0 \simeq 0.01 \div 0.02$ at $\sqrt{s} \sim 2\ TeV$.

In the regge type models the value of S_0 depends crucially on the relative weights of contributions of multipomeron exchanges to the elastic amplitude, and, in principle, can be extracted from "every" good model descriptions of $d\sigma/dt$ data at low t. But it is essential that in the most popular models one can not expect the functional behavior $S_0(y) \sim \exp(-yc)$ at large y, which seams very natural in the parton approach.

If $\tilde{v}(y, b)$ is the single $\mathbb{P}$ contribution to the amplitude, then the full sum of all pomeron diagrams (neglecting the pomeron interactions with one another) can be approximately

[7] Such type of model was proposed [4] long ago, and in this case (with large $\delta_y \sim 5$) it provides a simple explanation of unnaturally small values of some pomeron parameters, such as $\alpha'_{\mathbb{P}}$, $\Delta_{\mathbb{P}}$, $r_{3\mathbb{P}}, \ldots$

represented as some function of v in the form

$$\widetilde{S}(y,b) = \widetilde{S}[\tilde{v}] = \sum_{n=0}^{\infty} \frac{\gamma_n}{n!} (i\tilde{v})^n , \qquad \gamma_0 = \gamma_1 = 1 , \tag{8}$$

where the quantities $\gamma_n \geq 1$ for $n \geq 2$, and their values reflect the weights with which the diffractive jets contribute to the corresponding multipomeron vertices N_n.

In the simplest eikonal case (no inelastic diffraction !), when all $\gamma_n = 1$, we have $\widetilde{S} = \exp(i\tilde{v})$ and it leads to

$$|\widetilde{S}(y,b)| = \exp(-\ \mathrm{Im}\,\tilde{v}(y,b)) \sim \exp(-c_1 \exp(\Delta y)\ ) \tag{9}$$

Although such an eikonal-type amplitude (especially if properly adjusted [8]) can lead to a reasonable description of many data, there can be the incosistency with parton picture at very large y in any case.

The purely exponential behavior of $\widetilde{S}(y,b)$ can take place only if the eikonal coefficients γ_n grow like $n!$ at large n. In this case we have

$$|\widetilde{S}(y,b)| \simeq (c + \ \mathrm{Im}\,\tilde{v}(y,b))^{-1} \sim \exp(-\Delta y) \tag{10}$$

at large y.

For a reasonable behavior of series (8) the coefficients γ_n considered as function of n should be analytic for $\mathrm{Re}\, n \geq 0$. The asymptotics of $S[\tilde{v}]$ for $v \to \infty$ is determined by the most right singularities of γ_n in n. To have the asymptotics (10) the function γ_n must have the most right singularity at $n = -1$. The fairly interesting case corresponds to the behavior $\gamma_n = \Gamma(n+1)$ which corresponds to $c = 1$ in (10).

Note that such a grow of multi - $\mathbb{P}$ contributions, in comparison with the eikonal case, can essentially affect the form of the tail of multiplicity distribution. In the eikonal case the behavior of the tail of multiplicity distribution is roughly $\sigma_n \sim \sigma_{inel}/\Gamma(1 + n/\bar{n})$. In the case (10) the tail is more flat: $\sigma_n \sim \sigma_{inel} \exp(-cn/\bar{n})$. Note that experimentally the tail has the similar form $\sigma_n \sim \exp(-c_1 n/\bar{n})$, $c_1 \simeq 1.5$, where $\bar{n}$ is the mean multiplicity.

The experimental data on the behavior of $d\sigma/dt$ at high $|t|$ are not rich, especially at high s. The old data [3] on pp show the universal behavior of $d\sigma/dt \simeq 0.1\,|t|^{-8}$ for $2 \lesssim |t| \lesssim 15$ GeV^2 in the energy range $\sqrt{s} \simeq 30 \div 60$ GeV. At such energies the S_0 is still rather large, and the decrease of S_0 with s can in fact be compensated by the small growth (3) of $d\hat{\sigma}/dt$. New data at higher s and $|t|$ are needed for better understanding the behavior of S_0.

5. Conclusion

We end with a few remarks.

- The value of elastic S-matrix at zero impact parameter $S_0(y) = |\widetilde{S}(y, b = 0)|$ gives a probability that the fast hadron with energy $\sim m \exp(y)$ has no soft parton cloud.
  This quantity can be extracted from the behavior of elastic cross section $d\sigma/dt$ at relatively low $|t|$ (see Section 4).
  The same quantity also enters as a factor in the expression (2) for the elastic cross section $d\sigma/dt$ at relatively large $|t| > 2 \div 5 GeV^2$ (far outside of the diffraction peak), because hard scattering occurs mainly in states with a minimal number of partons.

[8] The "minimal" form of corrections to the simple eikonal is the quasi-eikonal picture, in which one explicitly introduces the bare parton state. A bit better and more consistent are the multichannel (matrix) eikonal models, but they also lead to the same rapid decrease of $\widetilde{S}$ as in (9).

- The existing data on $d\sigma/dt$ show that the value of $S_0(y)$ is not so small, as one can expect at first sight from the large multiplicity of secondary hadrons at the same energies. It indicates that the high energy hadron wave function contains a relatively small number of partons in average. This also means that most of secondary soft hadrons are created only after collision (they can come from the decay of nonperturbative QCD tubes or minijets), and so are not directly represented by the corresponding degrees of freedom in the incoming parton wave function. In most models of multiperipheral type the opposed picture is usually supposed. Therefore all new data on elastic $d\sigma/dt$ are very interesting also from this point of view, especially at maximally large (LHC) energies.
- Comparison of two different approaches, one of which is based on the parton model and the other on the Regge theory to estimation of the $S_0(y)$ suggests that the pure eikonal-type unitarization (when $\widetilde{S}(y,b) = \exp(i\tilde{v})$) can be not suitable at very high energies. The most simple form, giving the same type of answer for S_0 as in the parton case, is given by the expression $\widetilde{S}(y,b) = (1+ic\tilde{v})^{-1}$, and it corresponds to the case when the eikonal coefficients grow ($\sim n!$) due to large contribution of diffractive jets in multipomeron vertices N_n .

ACKNOWLEDGMENTS

I thank K.G. Boreskov for his interest and many useful suggestions and I.N. Dremin for helpful conversation. I remember numerous discussions with A.B. Kaidalov on various topics of high energy physics, including considered in this paper. These discussions were always interesting and very useful.

[1] S.J.Brodsky and G.R.Farrar Phys.Rev.Lett 31 (1973) 1153 ,
V.A.Matveev, R.M.Muradyan and A.N.Tavkhelidze, Lett.Nuovo Cim. 7 (1973)719
[2] G.Sterman, arXiv:1008.4122 ;
M.G.Sotiropoulos and G.Sterman, hep-ph/9401237 , Nuclear Physics **B**425 (1994) 489
[3] A. Donnachie and P.V. Landshoff, *Nucl.Phys.* **B244** (1984) 322;
[4] V.A.Abramovsky and O.V.Kancheli, Pisma Zh.Eksp.Teor.Fiz.32:498-501,1980.
[5] A.Capella, E.G.Ferreiro, C.A.Salgado, A.B.Kaidalov, Nucl.Phys. B593 (2001) 336,
Phys.Rev. D63 (2001) 054010
[6] J.Polchinski and M.J.Strassler, JHEP 0305:012,2003 ,
R.C.Brower, J.Polchinski, M.J.Strassler, C.I.Tan, JHEP 0712:005,2007

Journal of Physics: Conference Series **607** (2015) 012004 doi:10.1088/1742-6596/607/1/012004

Pomeron and its ups and downs

(My personal point of view)

Eugene Levin

Department of Particle Physics, School of Physics and Astronomy, Tel Aviv University, Tel Aviv, 69978, Israel
and
Departamento de Física, Universidad Técnica Federico Santa María and Centro Científico-Tecnológico de Valparaíso, Casilla 110-V, Valparaiso, Chile

E-mail: leving@post.tau.ac.il, eugeny.levin@usm.cl

Abstract. In this talk we review the main ideas on the Pomeron structure that have been developed during the half of century. Doing this I believe that I pay the tribute to my life long unfailing friend Aliosha Kaidalov who, as well as me, belongs to the Pomeron generation of high energy theoretical physicists.

1. Introduction

Aliosha Kaidalov, as well as me, belongs to Pomeron generation of high energy physicists. The structure of the Pomeron was, is and, I hope, will be the topic that has inspired us, exited us and, may be, even spoilt us for half of century. Therefore, I believe that presenting a brief review of our achievements and failures I pay my tribute to the memory of Aliosha,

The talk is organized in the following way. First, I give a brief history of Pomeron as a projection on my life in physics. I hope that you would feel the pulse of hopes and disappointments in the dry language of a scientific talk. Second, I share with you the four ideas on the Pomeron structure that made me happy, After this I give very personal and short review of the Pomeron that arises from high energy phenomenology, Finally I describe my the last paper on the subject of modeling confinement in the BFKL Pomeron.

2. Fifty years with the Pomeron

2.1. 1966 - 1974 ⟶ Frontier of high energy physics

The problem of finding of high energy asymptotic behaviour of the scattering amplitude was one of the priority problem of the particle physics. At that time we believed that the future theory will be based on three principles: analyticity, crossing symmetry and the unitarity constrains in t and s channels. In the framework of this approach the high energy asymptotic behaviour was needed to specify the use of the dispersion relation: our main tool to implement the above principles. We started to do research when the main approach to high energy amplitude at high energy had been formulated and it was based on the Reggeon approach. The theory

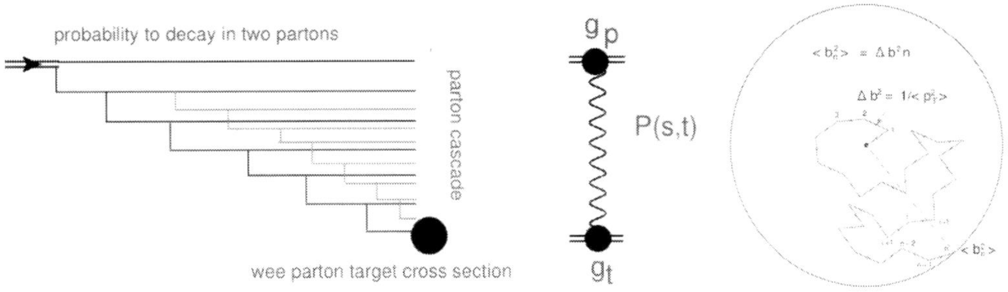

Fig. 1-a Fig. 1-b

Figure 1. The parton approach to high energy scattering: longitudinal structure of the parton cascade (Fig. 1-a) and Gribov's diffusion (Fig. 1-b).

of the Reggeon exchange had been developed but our generation solved two problems: the contribution the multi-Pomeron exchanges and the s-channel structure of the Pomeron exchange. Remembering that time I surprise how much we learned having so poor theoretical tool in hand. I believe that it happened because we were familiar with existing quantum field theories such as QED and other theoretical models but also because we had feedback interaction with the experimental physicists. We were searching not only how our approach describes the data but also what features in the data have general character that have to be implemented in the theoretical approach.

Basically we have very simple approach [1, 2, 3] based on the parton model, which can be summarized in two pictures of Fig. 1.

Fig. 1-a states that the total cross section for the single Pomeron exchange is equal to

$$\sigma = \sum_{n=2}^{\infty} \int_0^Y dy_1 \int_0^{y_1} dy_2 \ldots \int_0^{y_{n-1}} dy_n \prod_{i=2}^{n-1} d^2 p_{i,T}$$
$$\Psi^* \left(\{x_i, \vec{p}_{i,T}\}, y_n, \vec{p}_{n,T}\right) \Psi \left(\{x_i, \vec{p}_{i,T}, \}, \vec{p}_{n,T}\right) \sigma_{\mathrm{parton}} \left(y+n, p_{n,T}\right) \tag{1}$$

where Ψ is the wave function of the partons (point-like particles) which have restricted transverse momentum $p_{i,T} \leq \mu$ and μ does not depend on total energy. We assume that partons are distributed uniformly in the rapidity range $(0, Y)$ and the integral over dy_n converges.

Fig. 1-b illustrates the Gribov's diffusion picture in the transverse plane of the partons in the parton cascade. It based on the uncertainty principle that $\Delta b\, p_{i,T} \sim 1$ for emission of the parton in the cascade. This figure shows that after n-emissions the partons are distributed in the area with radius $b_n^2 = (1/\mu^2)\, n$ and since $n \propto Y$ we get $R^2 \propto (1/\mu^2)\, Y = \alpha'_{I\!P}\, Y$.

Based on this simple picture we found out how to calculate (i) the two particle correlations (Mueller diagrams [4, 5] see Fig. 2-a), (ii) the contribution of the multi-Pomerton exchange to the correlations functions[7, 8] (see Fig. 2-b) and (iii) to the processes of multi-particle production (AGK cutting rules[6], see Fig. 3); (v) learned the origin and main properties of the large mass diffractive production (see Fig. 2-c) as well as the (vi) structure of maxima and minima in elastic scattering (see Fig. 2-d); (vii) the general properties of the multiplicity distributions (KNO scaling or $\sigma_n/\sigma_{in} = F\left(n/ < n >\right)$ have been explained [9](see Fig. 2-e) and so on.

Our knowledge of Pomeron interaction was summarized in the Gribov Pomeron calculus[10] which can be formulated as the following path integral:

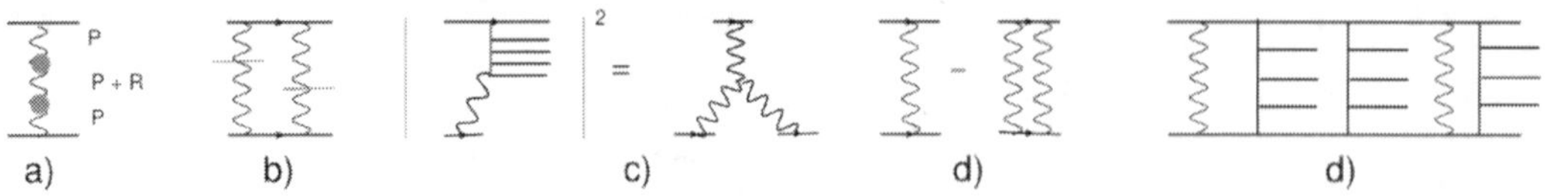

Figure 2. Diagrams for different processes: the two particle correlations (Fig. 2-a), two particles long rapidity range correlations (Fig. 2-b), the large mass diffractive production (Fig. 2-c), structure of maxima and minima in elastic scattering (Fig. 2-d) and KNO scaling (Fig. 2-e).

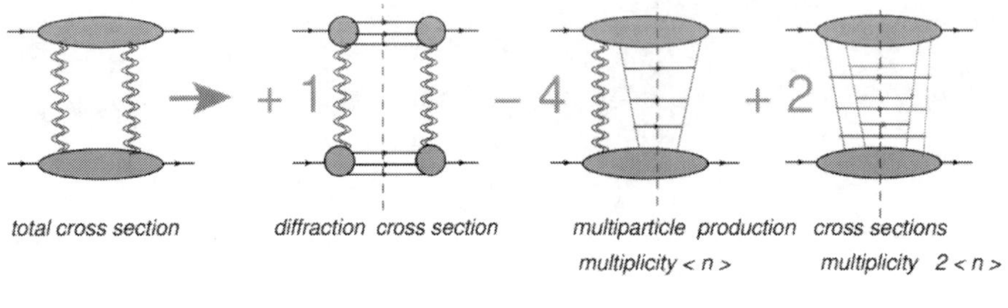

Figure 3. AGK cutting rules

$$Z[\Phi,\Phi^{+}] \,=\, \int D\Phi\, D\Phi^{+}\, e^{S} \text{ with } S \,=\, S_0 \,+\, S_I \,+\, S_E$$
$$S_0 \,=\, \int dY \Phi^{+}(Y,b)\left\{-\frac{d}{dY} \,+\, \Delta \,+\, \alpha'_{I\!P}\nabla^2\right\}\Phi(Y,b);$$
$$S_I \,=\, G_{3I\!P}\int dY\,\left\{\Phi(Y,b)\,\Phi^{+}(Y,b)\,\Phi^{+}(Y,b) \,+\, h.c.\right\} \tag{2}$$

S_E specifies the interaction with hadrons or nuclei. $\Phi\,(Y,r)$ describes the Pomeron with rapidity Y and impact parameter b. It turns out that Eq. (2) has a simple statistical interpretation and can be re-written as the equation for probability P_n to have n-Pomerons at rapidity Y. The equation takes the form[11, 12]

$$-\frac{\partial\, P_n(y,b)}{\partial\, y} \,+\, \alpha'_{I\!P}\nabla^2 P_n(y,b) \,=\, G_{3I\!P}\left\{-\,n\,P_n(y,b) \,+\, (n-1)\,P_{n-1}(y,b)\right\} \,+\, G_{3I\!P}\left\{-\,n\,(n-1)\,P_n(y,b) \,+\, (n+1)\,n\,P_{n+1}(y,b)\right\} \tag{3}$$

However, the problem of Pomeron interactions has not been solved and we failed to find theoretical arguments for restricting the number for Pomerton interaction vertices as well as for finding term S_E in Eq. (2).

2.2. 1974 - 1990 ⟶ Dark ages

In 1974 QCD was suggested and the theory of strong interaction got the solid theoretical basis. At first sight, both the idea of the Pomeron and its importance have no roots in QCD and they have been well forgotten at least for 15 years. During these years I had only a couple of publications on Pomeron. However, the problem of high energy amplitude in QCD was and is one of the most difficult problems which has been tried to solve during these years. The main ingredients of all attempt to approach this problem is so called BFKL Pomeron which is the solution to the BFKL evolution equation[13]. The main features of the BFKL Pomeron looks as follows:

- Scattering amplitude behaves as $s^{\Delta_{BFKL}}$ with $\Delta_{BFKL} = 4\ln 2\,\alpha_S$;
- σ Eq. (1) but the cross section $\sigma_{\rm parton}(y_n, p_{n,T})$ is a constant versus y;
- No Gribov's diffusion, $< b^2 >=$ Const;
- Diffusion in $\ln k_T^2$;
- The BFKL Pomeron interaction can be described by BFKL Pomeron calculus with the action which is very close to Eq. (2)[14];

By now the theoretical approach based on the BFKL Pomeron, has reached a mature stage (see Ref.[15]), we learned a lot about the high energy asymptotic behaviour of the scattering amplitude for deep inelastic and other processes. However, the soft scattering has been still elusive.

2.3. 1991 - present $\longrightarrow$ Renaissance

In 1991 I came to DESY for long visit and to my great surprise I noted, after staying six months, that the majority of questions that I was asked, ware related to the high energy Regge phenomenology. Frankly speaking, I had forgot what was a status of this phenomenology. Fortunately, Asher Gotsman, was also visiting DESY and stayed in the next door room. We started to clear up the situation and from that time A. Gotsman, U.Maor and me worked on the Regge phenomenology for soft interaction at high energy. Below, I will discuss what kind of approach we developed. During the past two decades I have followed closely the theoretical development of the understanding of the Pomeron structure and I will share with you four ideas on this subject that made me happy.

3. Four theoretical approaches that made me happy

3.1. Scaling anomaly (breakdown of scale invariance)

As we have mentioned perturbative QCD leads to the BFKL Pomeron after summing all diagrams that have contribution:$(\alpha_S\, lns)^n$. Generally speaking we expect that non-perturbative corrections stems from the region of low transverse momenta of gluon (Λ_{QCD}). In Ref.[16] we made an observation that strong gluon fields in QCD vacuum lead to sufficiently large produced mass in the BFKL ladder (see the second diagram of Fig. 4).

Indeed, due to scaling anomaly, $\alpha_S F^{\mu\nu,a}\, F_{\mu\nu,a} \;\propto\; \theta^\mu_\mu \;\sim\; \alpha^0_S$. Therefore, in the diagrams of Fig. 4 with two produced gluon in the rank of the ladder we see that contribution to the high energy asymptotic behaviour is proportional to $\ln^n s$, Using $< \pi^+\pi^-|\theta^\mu_\mu|0 >= M^2$ from chiral Lagrangian. Using the sum rules for θ^μ_μ [17]

$$\Pi\,(0) \;=\; i\int dx\,\langle 0|T\,\left\{\theta^\mu_\mu(x)\,\theta^\mu_\mu(0)\right\}|0\rangle \;=\; -\,4\langle 0|\theta^\mu_\mu|0\rangle \;=\; -16\epsilon_{vac}\,\neq\,0 \qquad (4)$$

Plugging in $< \pi^+\pi^-|\theta^\mu_\mu|0 >= M^2$ and subtraction perturbative contribution we obtain that

$$\Pi(0) \;=\; \int\frac{dM^2}{M^2}\,[\rho^{\rm phys}_\theta(M^2) - \rho^{\rm pt}_\theta(M^2)] = 16\,|\epsilon_{vac}| \qquad (5)$$

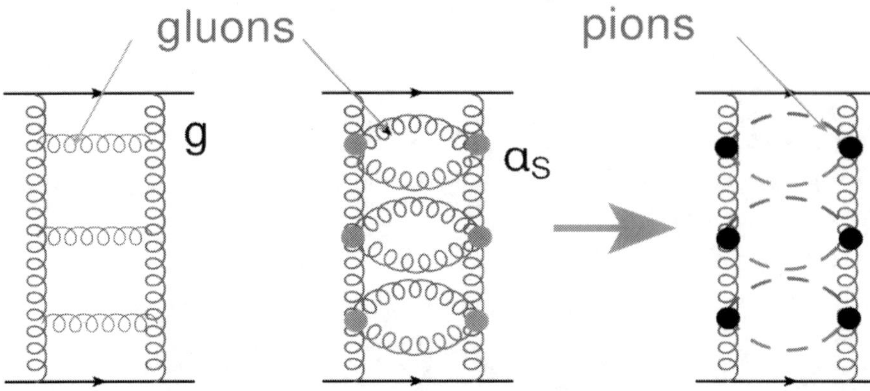

Figure 4. The diagrams for the BFKL Pomeron including the scale invariance breakdown

And using

$$\rho_\theta^{\rm phys}(M^2) \quad \xrightarrow{M^2 > M_0^2} \quad \rho_\theta^{\rm pt}(M^2) \tag{6}$$

we obtain that

$$M_0^2 \simeq 32\pi \left\{ \frac{|\epsilon_{\rm vac}|}{N_f^2 - 1} \right\}^{\frac{1}{2}} \tag{7}$$

Numerically: $\epsilon_{vac} \approx -(0.24\, GeV)^4$ and $M_0^2 = 4 \div 6\, GeV^2$ which gives the following intercept

$$\Delta = \frac{1}{48} \ln \frac{M_0^2}{4m_\pi^2} = 0.082 \div 0.09 \tag{8}$$

For me this was the first time were we obtain reasonable Pomeron intercept based on non-perturbative approach.

3.2. *Instantons and soft Pomeron*

In previous subsection we showed that the strong semi-classical field can play a crucial role in the soft Pomeron structure. Strong fields correspond to classical solution of QCD and there exists a well known such a solution: instantons (see Ref.[18] and reference therein). In our paper[19] we build the soft Pomeron from instantons using the simple multi-peripheral approach (see Fig. 5).

InFig. 6 we plotted the soft Pomeron trajectory as a function of $t\rho_0$ where ρ_0 is the typical scale of the QCD instanton.

Therefore, we demonstrated again that the strong gluon field can be dominant contribution to the soft Pomeron structure.

3.3. *AdS-CFT correspondence and the Pomeron in N=4 SYM*

In 70's Prof. Gribov, working on the Pomeron interactions in framework of the Pomeron calculus, told us, the young scientists in his department, that he needs a particle the same as photon but with vacuum quantum numbers. Indeed, his approach led to weak Pomeron coupling [21] and universality:$\Delta_{I\!P} = 0$; $g_h =$ Universal constant, $G_{3I\!P} = 0$.

More than decade ago such particle have been found. It turns out that due to AdS-CFT correspondence[20] the theory that can be solved in the limit of large coupling constant has

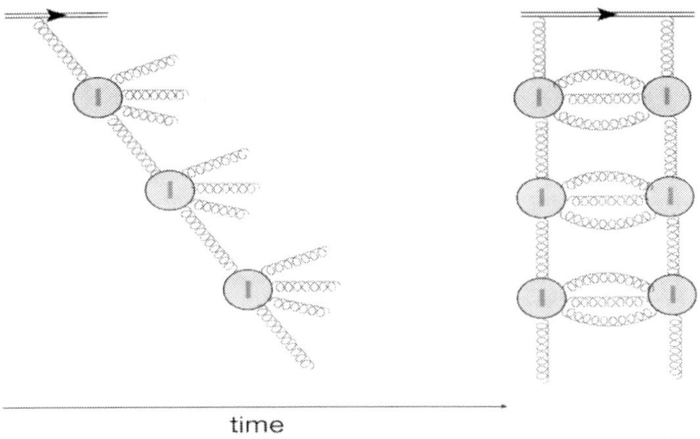

Figure 5. Instanton contribution to the soft Pomeron

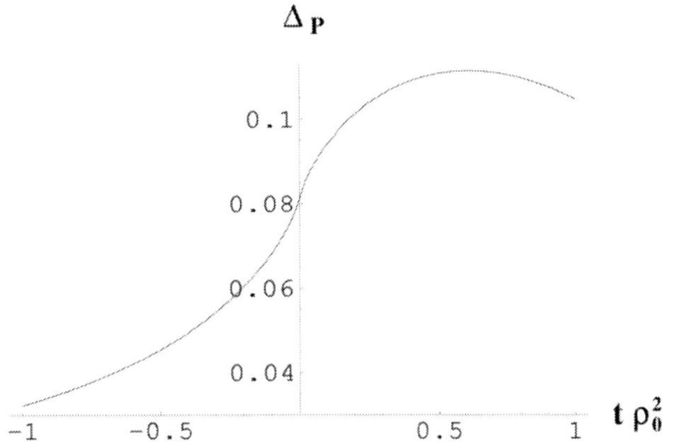

Figure 6. The soft Pomeron trajectory , as a function of $t\rho_0$, where ρ_0 is the typical scale of the QCD instanton, from Ref.[19].

been suggested. This theory: N=4 SYM, in the limit of small coupling has all typical property of QCD including the BFKL Pomeron. At large coupling this theory leads to the soft Pomeron exchange[22, 23]. It is instructive to notice that the Pomeron in this theory is not moving pole on the angular momentum, but the standing cut with all properties of the BFKL Pomeron.

Glossary $\equiv$ AdS-CFT correspondence:

N=4 SYM		QCD
Reggeized graviton	$\Longrightarrow$	BFKL Pomeron
z	$\Longrightarrow$	r (dipole size)
$1 - 2/\sqrt{\lambda}$	$\Longrightarrow$	ω_{BFKL}(intercept of BFKL Pomeron)
$2/\sqrt{\lambda}$	$\Longrightarrow$	$D_{BFKL}(\omega(\nu) \ = \ \omega_{BFKL} - D\nu^2)$

z is the fifth coordinate which has been introduced. The properties of the Pomeron in this approach look as follows.

$\lambda \gg 1$

$\Delta_{I\!P} \Longleftarrow$ large ($\sim$ 0.3?)

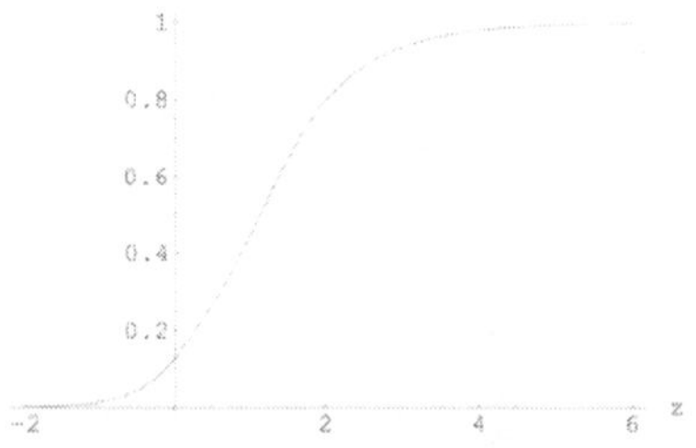

Picture is taken from Ref.[24].

$\alpha'_{I\!P} = 0$

only Good-Walker mechanism for diffraction

$G_{3I\!P} = 0$

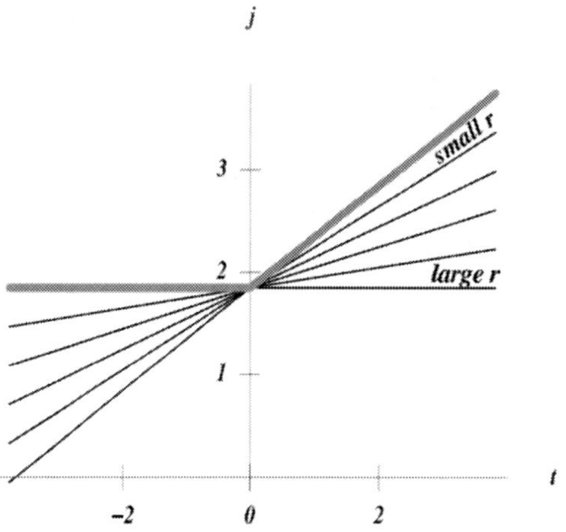

Picture is taken from Ref.[23]

Such Pomeron arises in top-down scenario in 10-dimensional de-Sitter space. However, even in bottom-up scenario (see Ref.[25] and references therein) where the string approach used in 5-dimensional space the resulting Pomeron is very similar to one described above.

In all approaches that we have discussed, is one major deficiency: they do not reproduce correct behavior at large impact parameter ($A \propto e^{-\mu b}$). This fact makes then interesting but questionable. Having this in mind we decided to build the high energy phenomenology which reproduce the main property of N=4 SYM Pomeron but has correct b dependence.

4. N=4 SYM and QCD motivated phenomenology

The presentation of our attempt to develop such phenomenology is given in the lecture of E.Gotsman [26]. Here I would like only to outline our approach to emphasize the fact that we tried to take into account the properties of the Pomeron both in N=4 SYM and in QCD. Building our model we assume:

- Pomeron is a Regge pole;
- $\Delta_{I\!P}$ is large $(0.2 \div 0.3)$;
- $\alpha'_{I\!P} = 0$;
- Large Good- Walker component, two channel model;

- Only $G_{3I\!P}$ as in QCD;
- $G_{3I\!P}$ is small (in QCD $G_{3I\!P} \propto \alpha_S$ and in N=4 SYM $G_{3I\!P} \ll 2/\sqrt{\lambda}$);

Our approach can be formulated in terms of functional integral

$$Z[\Phi, \Phi^+] = \int D\Phi\, D\Phi^+\, e^S \quad \text{with} \quad S = S_0 + S_I + S_E \tag{9}$$

Where S_0 describes the Pomeron without interaction. S_I is responsible for interaction between Pomerons while S_E is the term that describe the interaction of the Pomeron with the colliding particles. They are equal to

$$S_0 = \int dY \Phi^+(Y) \left\{ -\frac{d}{dY} + \Delta \right\} \Phi(Y); \tag{10}$$

$$S_I = G_{3I\!P} \int dY \left\{ \Phi(Y)\, \Phi^+(Y)\, \Phi^+(Y) + h.c. \right\}; \tag{11}$$

$$S_E = -\int dY \sum_{i=1}^{2} \left\{ \Phi(Y)\, g_i(b) + \Phi^+(Y)\, g_i(b) \right\}; \tag{12}$$

This model is so simple that can be solved. In Ref.[26] you can see that we achieved the good description of soft interaction from energy 20 GeV till LHC energy.

5. BFKL Pomeron: modeling confinement

5.1. Setting the problem

In this section we discuss briefly the results of our paper (see Ref.[27]) in which we address the question of large impact parameter dependence of the scattering amplitude. I hope, that I have discussed how important this question which has not been solved in all theoretical approaches to Pomeron, including BFKL Pomeron in perturbative QCD.

During the past decade numerous attempts (see our paper [27] for references) have been made to solve this problem without decisive result. However we learned several lessons from these tries: (i) the confinement of quarks and gluon have to be included in the BFKL kernel (to include in the initial conditions is not enough); (ii) suppressing large sizes of the produced dipoles in the decay of one dipole to two dipoles we reproduce correct b-dependence; and (iii) since at large b the amplitude is small we do not need to take into account the non-linear corrections.

Therefore, corrections from confinement have to be included in the kernel of the BFKL equation:

$$\frac{\partial N\left(x_{10}, b, Y\right)}{\partial Y} = \alpha_S \int d^2 x_{12} K\left(x_{12}, x_{20} | x_{10}\right) \times \left\{ 2\, N\left(x_{12}, \vec{b} - \frac{1}{2}\vec{x}_{02}; Y\right) - N\left(x_{10}, b; Y\right) \right\} \tag{13}$$

where kernel K describe the decay of one dipole to two dipoles ($x_{10} \to x_{12} + x_{02}$). We solve this equation with the modified BFKL kernel:

$$K\left(x_{12}, x_{20} | x_{10}\right) = \frac{x^2_{10}}{x^2_{12}\, x^2_{02}}\, e^{-B\left(x^2_{12} + x^2_{02}\right)} \tag{14}$$

The motivation for this behaviour of the wave function of a dipole in the confinement region stems from the Gaussian-like form of the wave functions of mesons in holographic AdS/QCD approach as well as in the phenomenology of the gluon emission at long distances. However, we will argue in conclusions that the main results of this paper do not depend on the particular form of Eq. (14).

5.2. *Summary of the results*

Since the lack of the room in this short version of the talk we start with the discussion of the main results of the paper.

(i) The scattering amplitude $N(x_{10}, b, Y) \xrightarrow{Bb^2 \gg 1} e^{-4Bb^2}$ (expected);
(ii) $N(x_{10}, b, Y) \xrightarrow{Y \gg 1} e^{\omega_0 Y}$ with $\omega_0 = \omega_{\rm BFKL}$ (unexpected !!!);
(iii) $\langle |b^2| \rangle$ = Constant (Y) (expected);
(iv) Saturation scale $Q_s^2 \propto e^{\lambda Y}$ $\lambda = \lambda_{\rm BFKL}$ (expected);
(v) The modified BFKL Pomeron looks similar to the Pomeron in N=4 SYM and in high energy phenomenology[26]: $\Delta_{I\!P} \sim 0.3$; $\alpha'_{I\!P} = 0$ (unexpected).

We postpone the discussion of the point 2 to the next section and briefly outline the results which were expected.

Large b behaviour of the amplitude. The large b behaviour of the amplitude, we can derive directly from Eq. (13) . Indeed, one can see that the main contribution at large b stems from the region where $|\vec{b} - \vec{x}_{12}| \le x_{10}$. At such x_{12} the equation takes the form

$$\frac{\partial N(x_{10} \approx 2b, b; Y)}{\partial Y} = \tag{15}$$

$$4\alpha_S \int \frac{d^2 x_{12}}{4b^2} e^{-4Bb^2 - Bx_{12}^2}\, 2\, N(x_{12}, 0; Y) \quad \propto \quad e^{-4Bb^2}$$

$\langle |b^2| \rangle$ dependence versus Y. The general origin of the energy dependence of $\langle |b^2| \rangle$ was found by Gribov[3] (Gribov's diffusion) and it stems from the uncertainty principle, that at each emission the shift in impact parameter (Δb) is proportional to $1/p_T$ where p_T is the average transverse momentum of emitted parton (gluon). In the parton model p_T is a dimensional scale of the model that does not depend on energy (rapidity) of the parton. In this case after n emission the average shift in b is equal to $\langle b^2 \rangle_n \propto \Delta b^2\, n$ and since $n \propto \ln s$ (s is the energy) we have the well known result that the interaction radius $R^2 \propto \ln s$. However, in QCD p_T increases with Y (energy) and after first several emission it becomes so large that we can neglect that shift in b. Our numerical calculation shows that, indeed, $\langle |b^2| \rangle$ is constant at large Y (see Fig. 7).

5.3. *The intercept of the modified BFKL Pomeron*

In this section we wish to discuss the main result of our approach: the intercept of the modified BFKL Pomeron turns out to be the same as the BFKL Pomeron. Going to ω-representation ($N\ (x_{01}; Y) = \int \frac{d\omega}{2\pi i} e^{\omega Y}\, N_\omega(x_{01})$) one can see that the modified BFKL equation takes the form

$$\omega N_\omega(x_{01}) = -\bar{\alpha}_S \mathcal{H}\, N_\omega(x_{01}) \tag{16}$$

or

$$E\, N_\omega(x_{01}) = \mathcal{H}\ N_\omega(x_{01}) \tag{17}$$

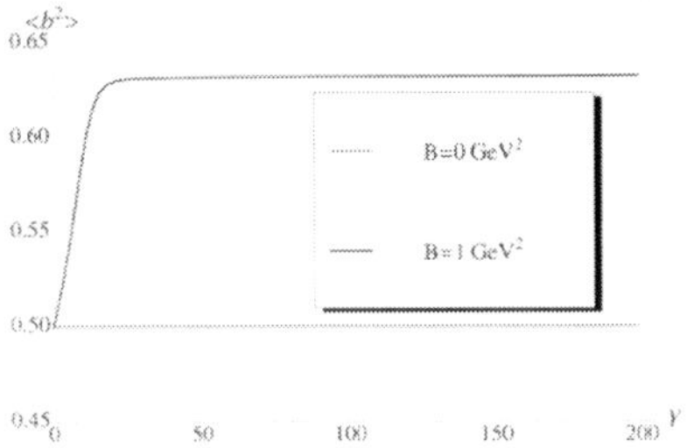

Figure 7. $\langle |b^2| \rangle$ versus Y for the modified (upper curve) and for the bare BFKL pomeron

For finding the eigenfunctions and eigenvalues of Eq. (17) we need to specify the boundary conditions. At short distances Eq. (14) has the same form as the BFKL kernel and, therefore, we have

$$N_\omega\left(x_{01}\right) \xrightarrow{x_{01}^2 \ll 1/B} N_\nu^{BFKL}\left(x_{01}\right) = \left(\frac{1}{x_{01}^2}\right)^{\frac{1}{2}+i\nu} \tag{18}$$

Note that the BFKL spectrum is

$$E\left(\nu\right) = 2\psi\left(1\right) - \psi\left(-\frac{1}{2} + i\nu\right) - \psi\left(\frac{1}{2} - i\nu\right) \tag{19}$$

For long distances the confinement modification of the kernel does not allow for dipoles to have large sizes leading to the boundary condition

$$N_\omega\left(x_{01}\right) \xrightarrow{x_{01}^2 \geq 1/B} \text{Constant} \tag{20}$$

Theory estimates: variational method. It is well known that the energy of the ground state is less or equal to

$$E_{\text{ground}} \equiv -\omega_0 \leq F[\{N\}] \tag{21}$$

where $F[\{N\}]$ is the following functional

$$F[\{N\}] = \frac{\left\langle N^*(x_{01}) \middle| \mathcal{H} \middle| N(x_{01}) \right\rangle}{\left\langle N^*(x_{01}) \middle| N(x_{01}) \right\rangle}$$

Choosing the set of the BFKL functions of Eq. (18) which is the complete and normalized set of functions, we find that $F_{min}[\{N\}] = \omega_{NFKL} = 4\ln 2\,\bar{\alpha_S}($ see Eq. (19) at $\nu = 0$ (see Fig. 8)

Therefore, the intercept of the modified BFKL equation could be only larger or equal to the intercept of the BFKL one.

In our paper [27] we developed two theoretical approaches: *semi-classical approach* and *diffusion approximation* which show that the modified BFKL equation does not have an intercept larger than the BFKL one. However, due to lack of room we cannot discuss these approaches here but they have been considered in our paper[27].

Numerical calculations.

Solving numerically we face two problems:

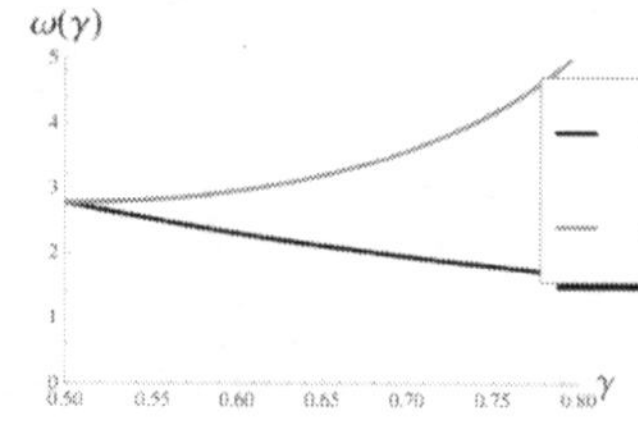

Figure 8. Comparison of $\omega(\gamma) = \frac{1}{2} + i\nu = F_{min}[\{N\}]$ with ω_{BFKL} given by Eq. (19).

(i) The kernel is not Fredholm type

$$\int d^2x_{01}d^2x_{12}\,K\left(x_{12},x_{02}|x_{01}\right) \implies \infty$$

(ii) The kernel is singular at $x_{12} \to x_{01}$

We use the following checks of our numerical procedure:

(i) Independence on the choice of x_{min} and x_{max};
(ii) Numerical solution to the BFKL equation coincide with the analytic one;
(iii) Independence on value of the regulator R;

$$\int d^2x_{13}K^B_R\left(x_{12},x_{02}|x_{10}\right)\,N\left(x_{12};Y\right) \equiv \tag{22}$$
$$\int d^2x_{12}\,\frac{e^{-B\left(x^2_{12}+x^2_{02}\right)}}{x^2_{12}+R^2}\left\{2\,N\left(x_{12};Y\right) - 2\,\frac{x^2_{10}}{x^2_{12}\,+\,x^2_{02}\,+\,2\,R^2}N\left(x_{10};Y\right)\right\}$$

(iv) Independence on value of B;

The result is plotted in Fig. 9.

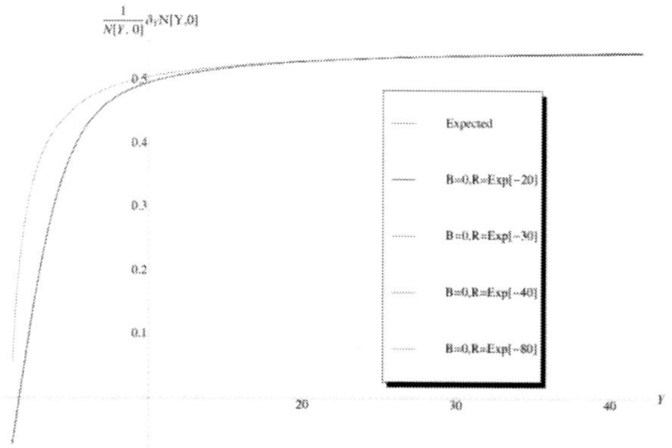

Figure 9. $d\ln N/dY$ versus Y for the BFKL (B=0) and modified BFKL (B=1) equations.

5.4. Conclusions

We found out that the modified BFKL Pomeron has the same intercept Δ as the BFKL Pomeron ($\Delta = \omega_{\rm BFKL} = 4\ln 2\bar{\alpha}_S$) and $\alpha'_{I\!P} = 0$. Therefore, the BFKL Pomeron with the modified kernel reproduces the main features of the soft Pomeron that has been found both from N=4 SYM theor and from the high energy Reggeon phenomenology.

Actually, we were surprised that the model for confinement changed so little in the BFKL Pomeron and on qualitative level, the Pomeron that emerges from the modified BFKL equation, looks quite the same at the BFKL Pomeron, both in parameters and in character of the energy behaviour. It seems that the only difference between the BFKL Pomeron and the modified BFKL Pomeron is that the second has a correct large impact parameter behaviour.

We believe that this statement does not depend on the particular form of Eq. (14). As we have mentioned the spectrum of the modified BFKL equation is determined by Eq. (17) with two boundary conditions of Eq. (18) at short distances and Eq. (20) at long distances. At first sight the condition at long distances will restrict the values of ν in the comparison with the BFKL equation. However, it is not the case. Indeed, the eigenvalues of the BFKL equation is degenerate having two eigenfunctions with positive and negative ν. One can see that we can find the sum of these two eigenfunction ($N(x_{12},\omega) = \left(1/x^2_{12}\right)^{\frac{1}{2}} \sin\left(\nu \ln\left(1/(B x^2_{12})\right) + \phi_0\right)$ which is equal to constant ($\sin\phi_0$) at $x_{12} = 1/B$. Therefore, at any ν we can satisfy the condition of Eq. (20). On the other hand for $\nu = i\kappa$ ($\kappa > 0$) we have two eigenfunctions: $\left(x^2_{12}\right)^{-\frac{1}{2}-\kappa}$ and $\left(x^2_{12}\right)^{-\frac{1}{2}+\kappa}$. The normalization condition selects out the only eigenfunction $\left(x^2_{12}\right)^{-\frac{1}{2}+\kappa}$ which has no divergency at $x_{12} \to 0$. Using this function we cannot satisfy the condition at $x_{12} \to \infty$ and, therefore, we have no solution of Eq. (17) for $\nu = i\kappa$. Hence, we expect that the spectrum for the modified Hamiltonian will be the same as the BFKL spectrum.

The independence of the spectrum of the BFKL Pomeron on the models for the confinement gives us a hope that the unknown confinement will change only slightly the equations of the CGC/saturation approach and these changes will not depend on the particular way of taking into account the long distances physics. In simple words, this paper gives a hope that the CGC/saturation approach will be still a theory in spite of needed model modifications due to confinement.

6. Resume

I wanted to demonstrate how great the progress which has been made in understanding of the Pomeron structure. However, the honest answer to the question: do we understand what is Pomeron, is no. The fact that the soft Pomeron appeared in AdS-CFT approach gives us some hope that the new generation of high energy physicists will find the answer.

References

[1] Feynman, R. P. 1969 *Phys. Rev. Lett.* **23** 1415; *"Photon-Hadron Interactions"*, Reading 1972, p282
[2] Gribov, V. N. 2000 *"Space-time description of hadron interactions at high energies"*, arXiv:hep-ph/0006158.
[3] Gribov, V. N. 1969 *Sov. J. Nucl. Phys.* **9** 369 [*Yad. Fiz.* **9** 640].
[4] Mueller, A. H. 1970 *Phys. Rev.* D **2** 963.
[5] Kancheli, O. V. 1970 *JETP Lett.* **11** 267 [*Pisma Zh. Eksp. Teor. Fiz.* **11** 397].
[6] Abramovsky, V. A. Gribov, V. N. and Kancheli, O. V. 1973 *Yad. Fiz.* **18**, 595 [*Sov. J. Nucl. Phys.* **18**, 308].
[7] Kaidalov A. B. and Karnakov, B. M. 1970 *Sov. J. Nucl. Phys.* **11** 121 [*Yad. Fiz.* **11** 216].
[8] Levin, E. M. and Ryskin, M. G. 1974 *Sov. J. Nucl. Phys.* **20** 280 [*Yad. Fiz.* **20** 519].
[9] Levin, E. M. and Ryskin, M. G. 1974 *Yad. Fiz.* **19**, 669
[10] Gribov, V. N. 1967 *Sov. Phys. JETP* **26** 414 [*Zh. Eksp. Teor. Fiz.* **53** 654].

[11] Grassberger, P. and Sundermeyer, K. 1978 *Phys. Lett.* **B77** 220.
[12] Boreskov, K.G. 2001 *"Probabilistic model of Reggeon field theory"*, arXiv:hep-ph/0112325 and reference therein.
[13] Kuraev, E. A. Lipatov, L. N. and Fadin, F. S. 1977 *Sov. Phys. JETP* **45**, 199 Balitsky, Ya. Ya. and Lipatov, L. N. 1978 *Sov. J. Nucl. Phys.* **28**, 22
[14] Braun, M. A. 2000 *Eur. Phys. J.* **C16** 337 [arXiv:hep-ph/0001268]; *Phys. Lett.* **B483** (2000) 115 [arXiv:hep-ph/0003004]; .
[15] Kovchegov , Y. V. and . Levin, E. 2012 "Quantum chromodynamics at high energy", CUP and references therein.
[16] Kharzeev, D. and Levin, E, 2000 *Nucl. Phys.* B **578**, 351 [hep-ph/9912216].
[17] Migdal, A.A. and Shifman, M.A. 1982 *Phys. Lett.* B**114** 445.
[18] Diakonov, D. 2003 *Prog. Part. Nucl. Phys.* **51**, 173 [hep-ph/0212026].
[19] Kharzeev, D. E. Kovchegov Y. V. and Levin, E. 2001 *Nucl. Phys.* A **690**, 621 [hep-ph/0007182].
[20] Maldacena, J. M. 1998 *Adv. Theor. Math. Phys.* **2** 231 [*Int. J. Theor. Phys.* **38** 1113] [arXiv:hep-th/9711200]; Gubser, S. S.. Klebanov, I. R and Polyakov, A. M. 1998 *Phys. Lett.* B **428** 105 [arXiv:hep-th/9802109]; Witten, E. 1998 *Adv. Theor. Math. Phys.* **2** 505 [arXiv:hep-th/9803131].
[21] Gribov, V. N. and Migdal, A. A. 1969 *Sov. J. Nucl. Phys.* **8**, 583 [*Yad. Fiz.* **8**, 1002].
[22] Kotikov, A. V. Lipatov, L. N. Onishchenko, A. I. and Velizhanin, V. N. 2004 *Phys. Lett.* B **595** 521 [Erratum-ibid. B **632** (2006) 754] [arXiv:hep-th/0404092]; Kotikov, A. V. Lipatov, L. N. 2000 *Nucl. Phys.* B **582** 19 [arXiv:hep-ph/0004008]. Kotikov, A. V. Lipatov, L. N. and Velizhanin, V. N. 2003 *Phys. Lett.* B **557** 114 [arXiv:hep-ph/0301021];
[23] Brower, R. C. Polchinski, J., Strassler, M. J. and Tan,C. I. 2007 *JHEP* **0712** 005 [arXiv:hep-th/0603115]; Brower, R. C. Strassler, M. J. and Tan,C. I. 2009 *JHEP* **0903**, 092 [arXiv:0710.4378 [hep-th]]; *JHEP* **0903**, 050 [arXiv:0707.2408 [hep-th]];
[24] Kotikov, A. V. talk in this proceedings.
[25] Basar, G., Kharzeev, D. E., Yee, H. -U. and Zahed, I 2012 *PoS QNP* **2012**, 116
[26] Gotsman, E. talk in this proceedings.
[27] Levin, E. and Tapia, S. 2013 *JHEP* **07** 183, arXiv:1304.8022 [hep-ph],

Elastic scattering of protons and their structure

I.M. Dremin

Lebedev Physical Institute, Moscow 119991, Russia

E-mail: `dremin@td.lpi.ru`

Abstract.
The experimental data about elastic scattering of protons and their theoretical interpretations are briefly reviewed. The time delay between the conference and preparation of the manuscript allowed me to modify it and include more recent important results. Some modern problems of elastic scattering of hadrons newly revealed by experimental data obtained at the LHC are discussed. Among them are the curious findings about the behavior of real and imaginary parts of the elastic scattering amplitude at non-forward scattering. The comparison with experiment at LHC energies shows that their ratio in this region can be drastically different from its values measured at low transferred momenta. Also, it is shown how the shape and the darkness of the interaction region of colliding protons change with increase of their energies. In particular, the collisions become fully absorptive at small impact parameters at LHC energies that results in some special features of inelastic processes. These problems ask for further careful treatment in a wide energy interval.

1. Few words about Alexei Kaidalov

Aliosha Kaidalov was a talented and generous person highly respected by physics community. He left important traces in high energy physics, especially, in the field of inelastic (in particular, diffractive) processes and elastic scattering of hadrons.

It is a honour for me to give a talk to his memory.

2. Introduction

Elastic scattering of protons is one of the sources to learn the strong interaction forces which were the main object of studies of Aliosha Kaidalov (e.g., see the papers [1, 2, 3]). This particular topic has been reviewed recently in my paper [4]. To shorten the presentation, I omit many details contained in there and presented during the talk but concentrate here on two problems which became quite actual nowadays. Namely, I'll discuss the behavior of the elastic scattering amplitude at non-forward direction and our knowledge of the shape and opacity of the interaction region of two colliding protons.

3. Elastic scattering

At the first sight, the experimental data about elastic scattering seem to look very simplified. The only information about this process consists of the measurement of the differential cross section of the process related to the scattering amplitude $f(s,t)$ in a following way

$$\frac{d\sigma}{dt} = |f(s,t)|^2. \tag{1}$$

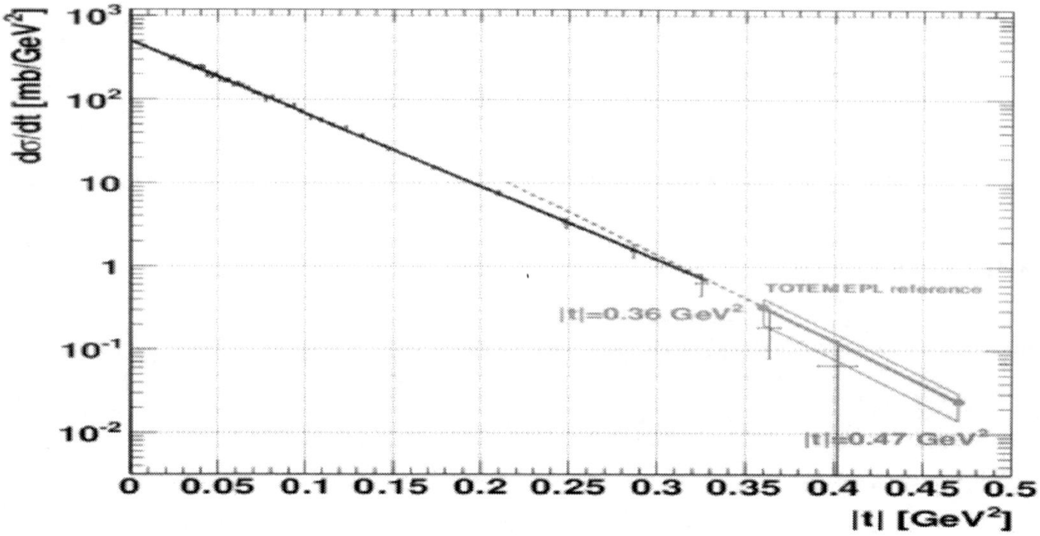

Fig. 1. The differential cross section of elastic proton-proton scattering at $\sqrt{s}$=7 TeV measured by the TOTEM collaboration [5].
The region of the diffraction cone with the $|t|$-exponential decrease is shown.

It is a function of two variables: $s = 4E^2$, where E is the energy of protons in the center of mass system, and the four-momentum transfer squared $-t = 2p^2(1 - \cos\theta)$ with θ denoting the scattering angle and p the momentum in the center of mass system. The amplitude f is normalized at $t = 0$ by the optical theorem such that

$$\mathrm{Im}f(s,0) = \sigma_t/\sqrt{16\pi}. \tag{2}$$

Moreover, one can find out the real part of the amplitude at very small t (practically at $t = 0$) from experimental data using the interference of the Coulomb and nuclear contributions to f. Even though the real and imaginary parts are, in general, related at any t by the dispersion relations as parts of a single analytic function, this treatment asks for some assumptions about their energy dependence, and the conclusions strongly depend on them.

Thus, from experiment, we get the knowledge only about the modulus of the amplitude at the available values of s and t and about the real and imaginary parts separately just in forward direction $t = 0$ but not at any other values of t. The theoretical approaches differ in ascribing different roles for their relative contributions at $t \neq 0$. Unfortunately, the available tools are rather moderate and can not exploit the power of QCD at full strength. Mostly, the phenomenological models and some insights from the unitarity relation are used.

The typical shapes of the differential cross section at small and larger values of $|t|$ are demonstrated in Figs 1 and 2. The most prominent feature of these plots is the fast decrease of the differential cross section with increasing transferred momentum $|t|$. As a first approximation at present energies, it can be described at comparatively small transferred momenta by the exponential shape with the slope B such that

$$\frac{d\sigma}{dt} \propto \exp(-B|t|). \tag{3}$$

This region is called the diffraction peak. Its slope B increases with energy approximately as $\ln s$. Moreover, it slightly depends on t at more careful fits of experimental data as seen in Fig.

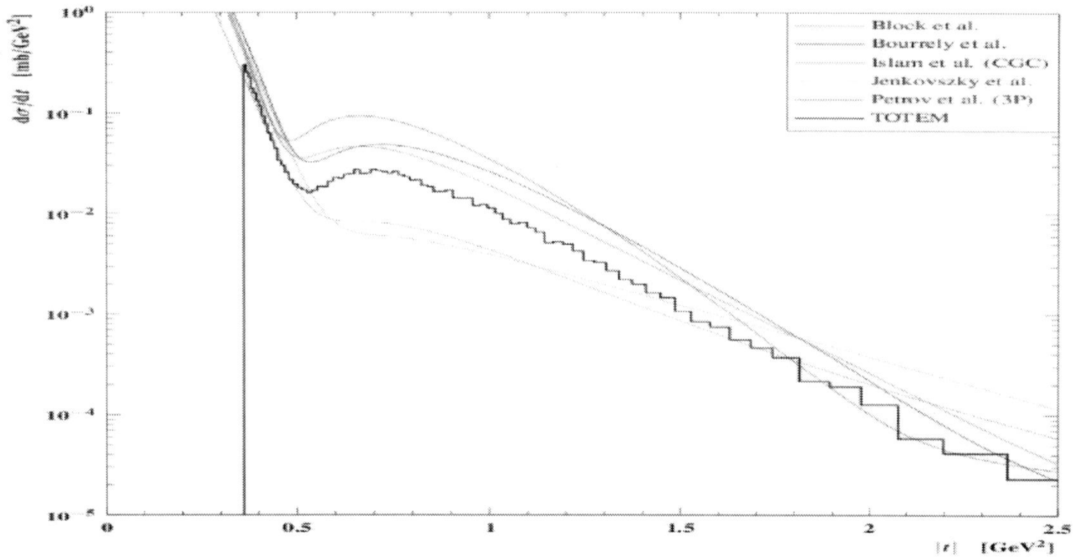

Fig. 2. The differential cross section of elastic proton-proton scattering at $\sqrt{s}$=7 TeV measured by the TOTEM collaboration [6].
The region beyond the diffraction peak is shown. The predictions of five models are demonstrated.

1. At larger values of $|t|$ outside the peak we observe the dip and slower decrease of the plots (see Fig. 2).

We also know from experiment the energy behavior of real and imaginary parts of the amplitude (or their ratio $\rho(s,0) = \rho_0$) in forward direction $t = 0$. At high energies it is rather small (equal to about 0.1 as measured at 7 TeV). Most phenomenological models try to fit namely these features in a wide energy range. Up to now, we can not claim that the desired aim has been achieved as seen, in particular, in Fig. 2 where the failure of predictions of five theoretical models is shown. This situation is described in more details in the review paper [4]. Actually, my talk at the conference followed it quite closely and I will not repeat it here.

Instead, I concentrate on the following two important problems. First, we discuss what we can say about the behavior of real and imaginary parts of the elastic scattering amplitude at any value of the transferred momentum and not only for forward scattering at $t = 0$. Second, we discuss what we can say about the space-time region of the interaction of the two high energy protons.

The first problem can be approached directly in the s, t-variables discussed above. It was shown a long ago [7, 8] that the imaginary part of the amplitude f outside the diffraction cone can be derived from the general unitarity condition which is reduced there to the linear integral equation

$$\mathrm{Im}f(p,\theta) = \frac{p\sigma_t}{4\pi\sqrt{2\pi B}} \int_{-\infty}^{+\infty} d\theta_1 \exp(-Bp^2(\theta-\theta_1)^2/2) g_\rho \mathrm{Im}f(p,\theta_1) + G(p,\theta), \tag{4}$$

where $g_\rho = 1 + \rho(s,0)\rho(s,\theta_1)$ and $G(p,\theta)$ is the overlap function.

Let us assume that the overlap function is negligible at these transferred momenta[1]. Then,

[1] The assumption of smallness of the overlap function outside the diffraction cone was shown to be valid [9, 11] by direct computation of it from experimental data in a wide energy interval.

the eigensolution of the homogeneous equation is

$$\mathrm{Im}f(p,\theta) = C_0 \exp\left(-\sqrt{2B\ln\frac{Z}{g_\rho}}p\theta\right) + \sum_{n=1}^{\infty} C_n \exp(-(\mathrm{Re}b_n)p\theta)\cos(|\mathrm{Im}b_n|p\theta - \phi_n) \tag{5}$$

with

$$Z = \frac{4\pi B}{\sigma_t}, \tag{6}$$

$$b_n \approx \sqrt{2\pi B|n|}(1 + i\mathrm{sign}n) \qquad n = \pm 1, \pm 2, ... \tag{7}$$

This expression contains the exponentially decreasing with θ (or $\sqrt{|t|}$) term (Orear regime!) with imposed on it oscillations strongly damped by their own exponential factors. These oscillating terms are responsible for the dip. Namely this formula was used in Refs [10, 11] for fits of experimental data in a wide energy range. The ratio ρ was approximated by its average values in and outside the diffraction cone so that $g_\rho = 1 + \rho_0\rho_l$ where ρ_l is treated as the average value of ρ in the Orear region. The fits at comparatively low energies [10] are consistent with $g_\rho \approx 1$, i.e., with small values of ρ_l close to zero. When $Z = 1$, as it happens at 7 TeV (see the Table below), the slope depends only on ρ and is very sensitive to its value. For the first time, that allowed to estimate the value of ρ in the Orear region at 7 TeV [11]. The great surprise of the fit of TOTEM data was necessity to use large in modulus negative value of $\rho_l \approx -2.1$ if $\rho_0 = 0.14$. It becomes larger in modulus $\rho_l \approx -3$ if the TOTEM value $\rho_0 = 0.1$ is used. Moreover, these values of ρ_l can be considered as upper limits because the effective value of ρ_0 inside the diffraction cone can be even smaller in view of its widely discussed zero there. No models have yet explained this finding. Further progress in solving the unitarity equation with proper dependence of ρ in and outside the diffraction cone is needed.

4. The structure function of proton

Let us come now to our second task. The structure of protons is one of the main problems in particle physics. It is successfully studied in electron-proton collisions. The point-like nature is ascribed to the colliding electron. The parton distribution functions of protons depending on the total energy and the virtuality of the exchanged photon are determined from experimental data. It is well known that they evolve with energy, i.e. the parton content of the proton evolves also. However, we do not get any knowledge about the space-time evolution of the interaction region.

In proton-proton collisions, both objects possess some complicated internal structure. The partons of one of them can interact with many partons from another one distributed somehow within some space volume. Therefore, it is hard to disentangle the individual contributions. Nevertheless, it becomes possible to study the space structure of the interaction region of the two protons formed by the combination of these effects. Here, we show what parameters obtained from experimental data about elastic scattering of protons influence such properties of this region as its size and opacity as well as its evolution with collision energy.

To define the geometry of the collision we must express all characteristics in terms of the transverse distance between the centers of the colliding protons called the impact parameter b. It is easily done by the Fourier – Bessel transform of the amplitude f written as

$$i\Gamma(s,b) = \frac{1}{2\sqrt{\pi}}\int_0^{\infty} d|t| f(s,t) J_0(b\sqrt{|t|}). \tag{8}$$

The unitarity condition in the b-representation reads

$$G(s,b) = 2\mathrm{Re}\Gamma(s,b) - |\Gamma(s,b)|^2. \tag{9}$$

The left-hand side (the overlap function) describes the impact-parameter profile of inelastic collisions of protons. It satisfies the inequalities $0 \leq G(s,b) \leq 1$ and determines how absorptive is the interaction region depending on the impact parameter (with $G = 1$ for full absorption). If integrated over the impact parameter, it leads to the cross section of inelastic processes. The terms on the right-hand side would produce the total cross section and the elastic cross section, respectively.

Also, the amplitude $f(s,t)$ may be connected to the eikonal phase $\delta(s,\mathbf{b})$ and to the opaqueness (or blackness $\Omega(s,\mathbf{b})$) at the impact parameter $\mathbf{b}$ by the Fourier – Bessel transformation

$$f(s,t=-q^2) = \frac{1}{2i\sqrt{\pi}}\int d^2b \exp(i\mathbf{qb})[\exp(2i\delta(s,\mathbf{b})-1] = \frac{i}{2\sqrt{\pi}}\int d^2b \exp(i\mathbf{qb})[1-\exp(-\Omega(s,\mathbf{b})]. \tag{10}$$

The integration is over the two-dimensional space of the impact parameter.

While the imaginary part of the amplitude in forward direction is defined by the relation (2), its real part can only be measured from interference with the Coulombic part of the amplitude. At high energies, its ratio to the imaginary part $\rho(s,0) = \rho_0$ happens to be rather small. Moreover, there are some theoretical arguments that this ratio is even smaller within the diffraction cone. The contribution from the diffraction peak prevails in Eq. (8). Therefore, in first approximation, the second term in Eq. (9) can be replaced by $[\mathrm{Re}\Gamma(s,b)]^2$. Respectively, one can assume that $\Omega(s,\mathbf{b})$ is real and use Eqs (2), (10) to get the following expressions for the total cross section

$$\sigma_t = 4\pi\int_0^R [1-\exp(-\Omega(s,\mathbf{b}))]bdb, \tag{11}$$

for the elastic cross section

$$\sigma_{el} = 2\pi\int_0^R [1-\exp(-\Omega(s,\mathbf{b}))]^2 bdb, \tag{12}$$

and for the inelastic cross section

$$\sigma_{in} = 2\pi\int_0^R [1-\exp(-2\Omega(s,\mathbf{b}))]bdb. \tag{13}$$

The black disk limit corresponds to $\Omega = \infty$, i.e. $\sigma_t = 2\sigma_{el} = 2\sigma_{in}$.

Using the above formulae, one can write the dimensionless Γ as

$$i\Gamma(s,b) = \frac{\sigma_t}{8\pi}\int_0^\infty d|t|\exp(-B|t|/2)(i+\rho(s,t))J_0(b\sqrt{|t|}). \tag{14}$$

Here, the diffraction cone approximation (3) is inserted. Herefrom, one calculates

$$\mathrm{Re}\Gamma(s,b) = \frac{1}{Z}\exp(-\frac{b^2}{2B}). \tag{15}$$

This dependence on the impact parameter was used, in particular, in [12]. The differential cross section is quite small outside the diffraction peak and does not influence the impact parameter profile G. Therefore, our second task happens to be practically independent of the first problem.

As was mentioned, the ratio $\rho(s,t)$ is very small at $t=0$ and, at the beginning, we neglect it and get

$$G(s,b) = \frac{2}{Z}\exp(-\frac{b^2}{2B}) - \frac{1}{Z^2}\exp(-\frac{b^2}{B}). \tag{16}$$

Table. The energy behavior of Z and $G(s,0)$.

$\sqrt{s}$, GeV	2.70	4.11	4.74	7.62	13.8	62.5	546	1800	7000
Z	0.64	1.02	1.09	1.34	1.45	1.50	1.20	1.08	1.00
$G(s,0)$	0.68	1.00	0.993	0.94	0.904	0.89	0.97	0.995	1.00

For central collisions with $b=0$ it gives

$$G(s,b=0)=\frac{2Z-1}{Z^2}. \tag{17}$$

Thus, the darkness of the central region is fully determined by the ratio Z. It becomes completely absorptive $[G(s,0)=1$ or $\Omega(s,0)=\infty]$ in the center only at $Z=1$ and diminishes for other values of Z. In the Table, we show the energy evolution of Z and $G(s,0)$ for pp and $p\bar{p}$ scattering.

The impact parameter distribution of the blackness $G(s,b)=1-\exp(-2\Omega(s,b))$ in Eq. (16) has the maximum at $b_m^2=-2B\ln Z$ with full absorption $G(b_m)=1$ or $\Omega(s,b_m)=\infty$. Its position depends both on B and Z. Note, that, for $Z>1$, one gets the incomplete absorption $G(s,b)<1$ at any physical $b\geq 0$ with the largest value reached at $b=0$ because the maximum appears at non-physical values of $b<0$. The disk is semi-transparent. At $Z=1$, the maximum is positioned exactly at $b=0$, and $G(s,0)=1$. The disk becomes black in the center. At $Z<1$, the maximum shifts to positive physical impact parameters. The dip is formed at the center that leads to the concave shape of the interaction region. It becomes deeper at smaller Z. The limiting value $Z=0.5$ leading to the complete transparency at the center $b=0$ is considered in more details below.

The maximum absorption in central collisions $G(s,0)=1$ is reached at the critical point $Z=1$ which is the case at $\sqrt{s}=7$ TeV considered first. Moreover, the strongly absorptive core of the interaction region grows in size as we see from expansion of Eq. (16) at small impact parameters:

$$G(s,b)=\frac{1}{Z^2}[2Z-1-\frac{b^2}{B}(Z-1)-\frac{b^4}{4B^2}(2-Z)]. \tag{18}$$

The second term proportional to b^2 vanishes at $Z=1$, and $G(b)$ develops a plateau which extends to quite large values of b about 0.4 - 0.5 fm. The plateau is very flat because the third term starts to play any role at 7 TeV (where $B\approx 20$ GeV^{-2}) only at even larger values of b. The structure of the interaction region with a central core at energies 7 - 8 TeV is also supported (see Fig. 3) by direct computation [13] using the experimental data of the TOTEM collaboration [5, 6] about the differential cross section in the region of $|t|\leq 2.5$ GeV^2. The results of analytical calculations and the computation practically coincide (see Fig. 1 in [14]). It was also shown in [14] that this two-component structure with the central black core and more transparent periphery is well fitted by the expression with the abrupt (Heaviside-like) change of the exponential. However, it is still pretty far from the black disk limit because the peripheral region at b near 1 fm is very active and shows strong increase compared to ISR energies [13]. The diffraction cone contributes mostly to $G(s,b)$. Therefore, the large-$|t|$ elastic scattering can not serve as an effective trigger of the black core even though some models were proposed (see, e.g., [15, 16]) which try to elaborate some predictions.

Inelastic exclusive processes can be effectively used for this purpose. One needs such triggers which enhance the contribution due to the central black core. Following the suggestions of [12, 17], it becomes possible [14] to study the details of the central core using the experimental data of CMS collaboration at 7 TeV about inelastic collisions with high multiplicity triggered by the jet production [18] as well as some other related data. Usage of very high multiplicity events

Journal of Physics: Conference Series **607** (2015) 012005 doi:10.1088/1742-6596/607/1/012005

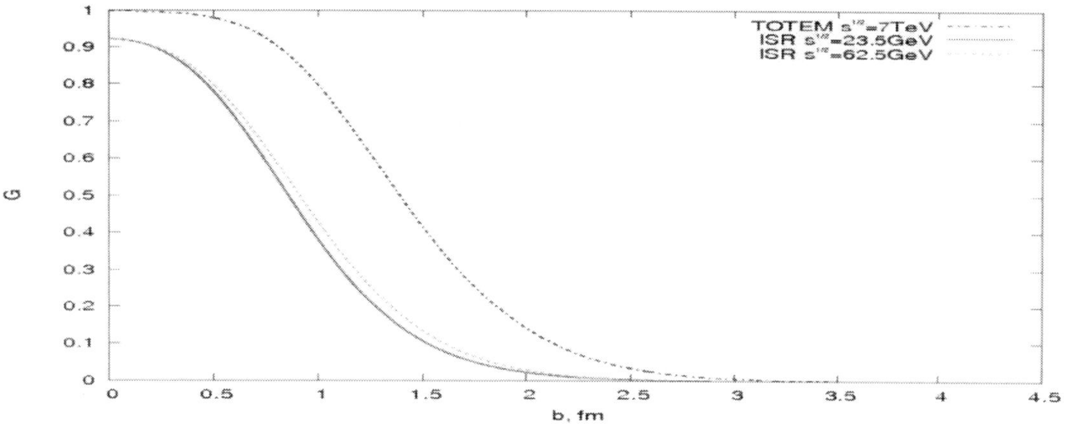

Fig. 3. The overlap function $G(s,b)$ at 7 TeV (upper curve) [11] compared to those at ISR energies 23.5 GeV and 62.5 GeV.

in combination with jet properties is crucial. Separating the core contribution with the help of these triggers, one comes to the important conclusion that the simple increase of the geometrical overlap area of the colliding protons does not account for properties of jet production at very high multiplicities. It looks as if the parton (gluon) density must strongly increase in central collisions and rare configurations (fluctuations) of the partonic structure of protons are involved.

It is interesting that the positivity of $G(s,b)$ imposes some limits on the relative role of B and σ_t. Namely, it follows from Eq. (17) that

$$2Z = \frac{8\pi B}{\sigma_t} \geq 1. \tag{19}$$

This relation implies that the slope B should increase asymptotically at least as strong as the total cross section σ_t. This inequality must be fulfilled even at intermediate energies.

It is usually stated that the equality $2Z = 8\pi B/\sigma_t = 1$ corresponds to the black disk limit with equal elastic and inelastic cross sections $\sigma_{el} = \sigma_{in} = 0.5\sigma_t$. However, one sees that $G(s,b=0) = 0$, i.e. the interaction region is completely transparent in central collisions. This paradox is resolved if we write the inelastic profile of the interaction region using Eq. (16). At $Z = 0.5$ it looks like

$$G(s,b) = 4[\exp(-\frac{b^2}{2B}) - \exp(-\frac{b^2}{B})]. \tag{20}$$

Recalling that $B = R^2/4$, we see that one should rename the black disk as a black torus (or a black ring) with full absorption $G(s,b_m) = 1$ at the impact parameter $b_m = R\sqrt{0.5\ln 2} \approx 0.59R$, complete transparency at $b = 0$ and rather large half-width about $0.7R$. Thus, the evolution to values of Z smaller than 1 at higher energies (this can happen if the decreasing tendency of Z with energy shown in the Table persists) would imply quite special transition from the two-scale features at the LHC to the concave torus-like configurations of the interaction region. The exponential shape of the diffraction cone at present energies must evolve to the typical for the black disk Bessel-like shape which is narrower than the exponent. The latter one acquires zero at quite small $|t| \approx 0.18$ GeV^2. No sign of it is seen at present energies beside a dip at $|t| = 0.53$ GeV^2 at 7 TeV. At the same time, the fit by a more steep exponential at the very edge of the diffraction peak used by TOTEM collaboration looks quite indicative to this tendency. The energy increase of the slope B (or the radius R) is also to be accounted. Further implications of this peculiar structure (if it appears!) for inelastic processes are to be guessed and studied.

In principle, the positivity of the inelastic cross section

$$\sigma_{in} = \frac{\pi B}{Z^2}(4Z - 1) \geq 0 \tag{21}$$

admits the value of Z as small as 0.25 which corresponds to $\sigma_{el} = \sigma_t$ and $\sigma_{in} = 0$. However, this possibility looks unphysical and has no interpretation in terms of eikonal (blackness).

Another consequence of Eq. (17) follows from study of energy evolution of $G(s,0)$ shown in the Table. In connection with the torus-like concave structure, it is interesting to point out the value of $Z = 0.64$ or $G(s,0) = 0.68$ at $\sqrt{s} = 2.70$ GeV and maximum 1 at $b_m^2 = 4B\ln 2$. One also notices that, in the energy interval 4 GeV$< \sqrt{s} < 8$ GeV, the values of Z are slightly larger than 1 so that the values of $G(s,0)$ are smaller but very close to 1. It looks as if the interaction region becomes black at the center $b = 0$ but at higher energies up to ISR loses this property trying to restore it at the LHC. This fact asks for further studies in the energy interval 4 GeV$< \sqrt{s} < 8$ GeV especially in view of proposed experiments in Protvino. The dark core must be smaller there than at LHC because of smaller values of B. Moreover, the contribution due to the real part of the amplitude is larger at these energies as well as larger $|t|$ beyond the diffraction cone can be important. One should also notice that Z becomes less than 1 at even smaller energies. As is easily shown, that does not pose any problem with the requirement $G(s,b) \leq 1$.

5. Conclusions

To conclude, we have shown that the absorption at the center of the interaction region of protons is determined by a single energy-dependent parameter Z. The region of full absorption extends to quite large impact parameters if Z tends to 1. This happens at $\sqrt{s} = 7$ TeV where the two-scale structure of the interaction region of protons becomes well pronounced. That leads to special consequences both for elastic and inelastic processes. The value of $Z = 1$ attained at the LHC energies is crucial for the behavior of the differential cross section outside the diffraction cone. Its slope there becomes fully defined by the ratio of the real part of the amplitude to its imaginary part yet unmeasured in this range of the transferred momenta. Analysis of experimental data at 7 TeV about the slope of the differential cross section inside the Orear region reveals that this ratio is negative and large in modulus. The behavior of Z at higher energies is especially important for the evolution of the geometry of the interaction region.

I am grateful for support by the RFBR grants 12-02-91504-CERN-a, 14-02-00099 and the RAS-CERN program.

[1] A.B. Kaidalov, Sov. J. Nucl. Phys. **13**, 401 (1971).
[2] A.B. Kaidalov, K.A. Ter-Martirosian, Phys. Lett. B **117**, 247 (1982).
[3] I.M. Dremin, A.B. Kaidalov, UFN **176**, 275 (2006); Physics-Uspekhi **49**, 263 (2006).
[4] I.M. Dremin, UFN **183**, 3 (2013); Physics-Uspekhi **56**, 3 (2013).
[5] G. Antchev et al. TOTEM Collaboration, Europhys. Lett. **95**, 41001 (2011).
[6] G. Antchev et al. TOTEM Collaboration, Europhys. Lett. **96**, 21002 (2011).
[7] I.V. Andreev, I.M. Dremin, ZhETF Pis'ma **6**, 810 (1967).
[8] I.V. Andreev, I.M. Dremin, Sov. J. Nucl. Phys. **8**, 473 (1969).
[9] I.V. Andreev, I.M. Dremin, D.N. Steinberg, Sov. J. Nucl. Phys. **11**, 261 (1970).
[10] I.V. Andreev, I.M. Dremin, I.M. Gramenitskii, Nucl. Phys. **10** 137 (1969).
[11] I.M. Dremin, V.A. Nechitailo, Phys. Rev. D **85**, 074009 (2012).
[12] L. Frankfurt, M. Strikman, C. Weiss, Phys. Rev. D **83**, 054012 (2004).
[13] I.M. Dremin, V.A. Nechitailo, Nucl. Phys. A **916**, 241 (2013).
[14] M.Yu. Azarkin, I.M. Dremin, M. Strikman, Phys. Lett. B (2014) ;arXiv:1401.1973.
[15] B.Z. Kopeliovich, I.K. Potashnikova, B. Povh, E. Predazzi, Phys. Rev. Lett. **85**, 507 (2000).
[16] M.M. Islam, R.J. Luddi, A.V. Prokudin, Mod. Phys. Lett. A **18**, 743 (2003).
[17] L. Frankfurt, M. Strikman, C. Weiss, Phys. Rev. D **69**, 114010 (2004).
[18] CMS Collaboration, Eur. Phys. J. C **73**, 2674 (2013).

Net-Baryon Production in Nucleus-Nucleus Collisions and Rare QGP Events*

G.H. Arakelyan[1], **C. Merino**[2], **and Yu.M. Shabelski**[3]

[1]A.Alikhanyan National Scientific Laboratory, Yerevan Physics Institut, Yerevan, 0036, Armenia

E-mail: argev@mail.yerphi.am

[2]Departamento de Física de Partículas, Facultade de Física, and Instituto Galego de Física de Altas Enerxías (IGFAE), Universidade de Santiago de Compostela, 15782 Santiago de Compostela, Galiza-Spain

E-mail: merino@fpaxp1.usc.es

[3]Petersburg Nuclear Physics Institute, NCR Kurchatov Institute, Gatchina, St.Petersburg 188350, Russia

E-mail: shabelsk@thd.pnpi.spb.ru

* *Talk given by C. Merino*

Abstract. The experimental data on net proton and net Λ-hyperon spectra obtained by the NA35 Collaboration, as well as the inclusive densities of Λ and $\overline{\Lambda}$ obtained by NA49, NA57, and STAR collaborations, are compared with the predictions of the Quark-Gluon String Model. The contributions of String Junction diffusion, interactions with nuclear clusters, and percolation corrections are accounted for. The level of numerical agreement of the calculations with the experimental data is of about 20−30%. We also consider the experimental ratios of multistrange to strange antibaryon production. The significant differences between the experimental $\overline{\Xi}^+/\overline{\Lambda}$ and, especially, $\overline{\Omega}^+/\overline{\Lambda}$ values and the QGSM predictions can be interpreted as the signal of the existence of an additional source of multistrange antibaryon production.

1. Introduction

The Quark-Gluon String Model (QGSM) [1] is based on the Dual Topological Unitarization, Regge phenomenology, and nonperturbative notions of QCD. This model is successfully used for the description of multiparticle production processes in hadron-hadron [2, 3, 4, 5], hadron-nucleus [6, 7], and nucleus-nucleus [8, 9, 10] collisions. In particular, the rapidity dependence of inclusive densities of different secondaries ($\pi^{\pm}$, $K^{\pm}$, p, and $\overline{p}$) produced in Pb-Pb collisions at 158 GeV/c per nucleon were reasonably described in reference [10].

In the QGSM high energy interactions are considered as proceeding via the exchange of one or several Pomerons, and all elastic and inelastic processes result from cutting through or between Pomerons [11]. Inclusive spectra of hadrons are related to the corresponding fragmentation functions of quarks and diquarks, which are constructed using the Reggeon counting rules [12].

In the case of interaction with nuclear target, the Multiple Scattering Theory (Gribov-Glauber Theory) is used. It allows to consider the interaction with nucleus as the superposition of interactions with different numbers of target nucleons [13, 14, 15, 16].

Also in the case of nucleus-nucleus collisions the Multiple Scattering Theory allows to consider the interaction as the superposition of separate nucleon-nucleon interactions. Though in this case the analytical summation of all the diagrams is impossible [17], the signiificant classes of diagrams can be analytically summed up in the so-called rigid target approximation [18] which is used in the present paper.

The significant differences in the yields of baryons and antibaryons in the central (midrapidity) region are present even at high energies. This effect can be explained [4, 5, 19, 20, 21, 22, 23, 24] in QGSM by the special structure of baryons consisting of three valence quarks together with a special configuration of gluon field, called String Junction [25, 26, 27, 28].

One additional contribution comes from the coherent interaction of a projectile with multiquark clusters inside the nuclei. The existence of these interactions is confirmed by the presence, with not such a small probability, of a cumulative effect [29]. These contributions were incorporated into the QGSM in [30, 31], and they allow to describe a number of experimental facts.

At very high energies the contribution of the enhancement Reggeon diagrams (percolation effects) becomes important, what leads to a new phenomenological effect, the suppression of the inclusive density of secondaries [32] into the average central (midrapidity) region. This corresponds to a significant fusion of the produced quark-gluon strings. In the energy limit when the probability of this fusions is very large, one should expect the appearance of a new state of matter, the Quark-Gluon Plasma(QGP). However, the process of QGP production can already occur with small probability at a not so high energy.

In this paper we present [33] the description of p, $\overline{p}$, Λ, and $\overline{\Lambda}$ production on nuclear targets at CERN SpS and RHIC energies. We also consider [34] the ratios of multistrange to strange antihyperon production in nucleon-nucleus and in nucleus-nucleus collisions.

Let us define, for the collision of projectil A (nucleon or nucleus) with a nuclear target B:

$$R(\overline{\Xi}^+/\overline{\Lambda}) = \frac{dn}{dy}(A+B\to\overline{\Xi}^+ + X)/\frac{dn}{dy}(A+B\to\overline{\Lambda}+X)\,, \tag{1}$$

$$R(\overline{\Omega}^+/\overline{\Lambda}) = \frac{dn}{dy}(A+B\to\overline{\Omega}^+ + X)/\frac{dn}{dy}(A+B\to\overline{\Lambda}+X)\,. \tag{2}$$

The produced antihyperons, $\overline{\Xi}^+$ and $\overline{\Omega}^+$, contain valence antiquarks newly produced during the collision. The ratios in equations 1 and 2 are reasonably described by QGSM in the cases when a not very large number of incident nucleons participate in the collision (nucleon-nucleus or peripheral nucleus-nucleus collisions).

The number of quark-gluon strings (cut pomerons) in nucleus-nucleus collisions increases with centrality. If the secondaries are produced independently in every quark-gluon string, the ratio of yields of different particles should not depend on centrality. However, experimentally the enhancement of the yield of $\overline{\Xi}^+$ is stronger than that of $\overline{\Lambda}$, and for $\Omega^- + \overline{\Omega}^+$ the enhancement is also stronger than for $\overline{\Xi}^+$ (e.g. see [35]). This means that an additional source of multistrange hyperons appears from collective interactions of several strings, and this additional source can be considered as a QGP state effect.

2. Baryon/antibaryon asymmetry in the QGSM

2.1. General approach

The QGSM [1, 2, 3] allows one to make quantitative predictions for different features of multiparticle production, in particular, for the inclusive densities of different secondaries, both in the central and in the beam fragmentation regions.

Journal of Physics: Conference Series **607** (2015) 012006 doi:10.1088/1742-6596/607/1/012006

In QGSM, each exchanged Pomeron corresponds to a cylindrical diagram, and thus, when cutting one Pomeron, two showers of secondaries are produced. The inclusive spectrum of a secondary hadron h is then determined by the convolution of the diquark, valence quark, and sea quark distributions, $u(x,n)$, in the incident particles, with the fragmentation functions, $G^h(z)$, of quarks and diquarks into the secondary hadron h. In the case of multipomeron exchange ($n > 1$), a new contribution from sea quarks-antiquarks appears. Both the distributions and the fragmentation functions are constructed using the Reggeon counting rules [12]. The average number of exchanged Pomerons $\langle n\rangle_{pp}$ slowly increases with the energy. The details of the model are presented in [1, 2, 3, 19]. The values of the Pomeron parameters have been taken from [3].

For a nucleon target, the inclusive rapidity, y, or Feynman-x, x_F, spectrum of a secondary hadron h has the form [1]:

$$\frac{dn}{dy} = \frac{x_E}{\sigma_{inel}} \cdot \frac{d\sigma}{dx_F} = \sum_{n=1}^{\infty} w_n \cdot \phi_n^h(x) + w_D \cdot \phi_D^h(x) \ , \tag{3}$$

where the functions $\phi_n^h(x)$ determine the contribution of diagrams with n cut Pomerons, w_n is the relative weight of this diagram, and the term $w_D \cdot \phi_D^h(x)$ accounts for the contribution of diffraction dissociation processes.

In the case of pp collisions:

$$\phi_n^h(x) = f_{qq}^h(x_+,n) \cdot f_q^h(x_-,n) + f_q^h(x_+,n) \cdot f_{qq}^h(x_-,n) + 2(n-1) \cdot f_s^h(x_+,n) \cdot f_s^h(x_-,n) \ \ , \tag{4}$$

$$x_\pm = \frac{1}{2}[\sqrt{4m_T^2/s + x^2} \pm x] \ \ , \tag{5}$$

where f_{qq}, f_q, and f_s correspond to the contributions of diquarks, valence quarks, and sea quarks, respectively.

These contributions are determined by the convolution of the diquark and quark distributions with the fragmentation functions, e.g.,

$$f_q^h(x_+,n) = \int_{x_+}^{1} u_q(x_1,n) \cdot G_q^h(x_+/x_1) dx_1 \ \ . \tag{6}$$

In the calculation of the inclusive spectra of secondaries produced in pA collisions we should consider the possibility of one or several Pomeron cuts in each of the ν blobs of proton-nucleon inelastic interactions. It is essential to take into account all diagrams with every possible Pomeron configuration and its corresponding permutations. The diquark and quark distributions and the fragmentation functions are the same as in the case of pN interaction.

The total number of exchanged Pomerons becomes as large as

$$\langle n\rangle_{pA} \sim \langle\nu\rangle_{pA} \cdot \langle n\rangle_{pN} \ , \tag{7}$$

where $\langle\nu\rangle_{pA}$ is the average number of inelastic collisions inside the nucleus (about 4 for heavy nuclei at SpS energies).

The condition that the absorptive parts of the hadron-nucleus amplitude are determined by the combination of the absorptive parts of the hadron-nucleon amplitudes is basically satisfied [13, 14, 15, 16].

In the case of a nucleus-nucleus collision, in the fragmentation region of projectile we use the approach [8, 9, 10], where the beam of independent nucleons of the projectile interact with the target nucleus, what corresponds to the rigid target approximation [18] of Glauber Theory. In the target fragmentation region, on the contrary, the beam of independent target nucleons interact with the projectile nucleus, these two results coinciding in the central region. The corrections for energy conservation play here a very important role if the initial energy is not very high. This approach was used in [10] for the succsessful description of $\pi^\pm$, $K^\pm$, p, and $\overline{p}$ produced in Pb-Pb collisions at 158 GeV per nucleon.

2.2. String Junction contribution

In the string models, baryons are considered as configurations consisting of three connected strings (related to three valence quarks), called String Junction (SJ) [25, 26, 27, 28], this picture leading to some quite general phenomenological predictions.

The production of a baryon-antibaryon pair in the central region usually occurs via SJ-$\overline{SJ}$ pair production (SJ has upper color indices, whereas anti-SJ ($\overline{SJ}$) has lower indices), which then combines with sea quarks and sea antiquarks into a $B\overline{B}$ pair [27, 36], as it is shown in figure 1a.

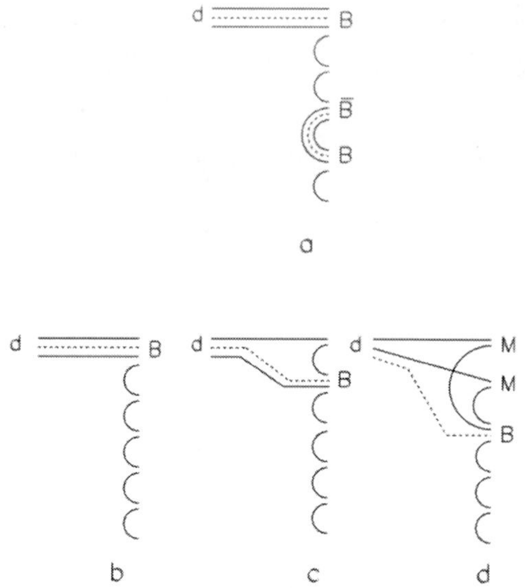

Figure 1. QGSM diagrams describing secondary baryon production: (a) usual $B\overline{B}$ central production with production of new SJ pair; (b) initial SJ together with two valence quarks and one sea quark; (c) initial SJ together with one valence quark and two sea quarks; (d) initial SJ together with three sea quarks.

However, in the processes with incident baryons another possibility to produce a secondary baryon in the central region, called SJ diffusion, exist. The quantitative description of the baryon number transfer due to SJ diffusion in rapidity space was obtained in [19] and following papers [4, 5, 20, 21, 22, 24].

In the QGSM the differences in the spectra of secondary baryons and antibaryons appear for processes which present SJ diffusion in rapidity space. These differences only vanish rather slowly when the energy increases.

To obtain the net baryon charge, and according to reference [19], we consider three different possibilities. The first one is the fragmentation of the diquark giving rise to a leading baryon (figure 1b). A second possibility is to produce a leading meson in the first break-up of the string and a baryon in a subsequent break-up [12, 37] (figure 1c). In these two first cases the baryon number transfer is possible only for short distances in rapidity. In the third case, shown in figure 1d, both initial valence quarks recombine with sea antiquarks into mesons, M, while a secondary baryon is formed by the SJ together with three sea quarks [5, 19, 38].

The fragmentation functions for the secondary baryon B production corresponding to the three processes shown in figures 1b, 1c, and 1d can be written as follows (see [19] for more details):

$$G^B_{qq}(z) = a_N \cdot v^B_{qq} \cdot z^{2.5} , \quad (8)$$
$$G^B_{qs}(z) = a_N \cdot v^B_{qs} \cdot z^2 \cdot (1 - z) , \quad (9)$$
$$G^B_{ss}(z) = a_N \cdot \varepsilon \cdot v^B_{ss} \cdot z^{1-\alpha_{SJ}} \cdot (1 - z)^2 , \quad (10)$$

where a_N is the normalization parameter, and v^B_{qq}, v^B_{qs}, v^B_{ss} are the relative probabilities for different baryons production that can be found by simple quark combinatorics [39, 40]. These probabilities depend on the strangeness suppression factor S/L, and we use S/L=0.32 following [41]. The contribution of the graph in figure 1d has in QGSM a coefficient ε which determines the small probability for such a baryon number transfer.

The fraction z of the incident baryon energy carried by the secondary baryon decreases from figure 1b to figure 1d. Only the processes in figure 1d can contribute to the inclusive spectra in the central region at high energies if the value of the intercept of the SJ exchange Regge-trajectory, α_{SJ}, is large enough. The analysis in [24] gives a value of $\alpha_{SJ} = 0.5 \pm 0.1$, that is in agreement with the ALICE Collaboration result, $\alpha_{SJ} \sim 0.5$ [42], obtained at LHC. In the calculations of these effects we use the following values of the parameters [24]:

$$\alpha_{SJ} = 0.5 \text{ and } \varepsilon = 0.0757 . \quad (11)$$

2.3. Contribution from interaction with clusters

In the case of interaction with a nuclear target some secondaries can be produced in the kinematical region forbidden for the interaction with a free nucleon. Such processes are called the cumulative ones, the simplest example being the production of secondary nucleons in the backward hemisphere in the laboratory frame.

Usually, the cumulative processes are considered as a result of the coherent interaction of a projectile with a multiquark cluster, i.e. with a group of several nucleons which are at short distances from each other that appears as as a fluctuation of the nuclear matter [29, 43].

The inclusive spectra of the secondary hadron h in the central region is determined at high energies by double-Pomeron diagrams [44]. The case of pp collision is shown in figure 2a. In the case of proton-nucleus collisions two different possibilities exist, the interactions with individual target nucleons (figure 2b) and the secondary production on cluster (figure 2c). For nucleus-nucleus collisions there are four possibilities (figures 2d, 2e, 2f, and 2g). The one in figure 2g corresponds to a new process where a secondary hadron is produced by the interaction of two clusters.

It was shown in references [30, 31] that in the case of secondary production from the cluster fragmentation, the inclusive spectra can be calculated in the framework of the QGSM with the same quark and diquark fragmentation functions. The only difference comes from the quark and diquark distributions, $u^{cl}_q(x, n, k)$ and $u^{cl}_{qq}(x, n, k)$, where k is the number of nucleons in the cluster. The distributions $u^{cl}_i(x, n, k)$ can also be calculated by using the Reggeon counting rules. All these functions $u^{cl}_i(x, n, k)$ are normalized to unity, and the function $u^{cl}_s(x, n)$ does not depend on k.

This approach was succesfully used in [30, 31] for the description of cumulative particles produced in hA collisions. In the present paper we use it for describing the enhancement of strangeness production on nuclear targets in the central region.

2.4. Inelastic screening (percolation) effects

The QGSM gives a reasonable description [5, 6, 10, 45] of the inclusive spectra of different secondaries produced both in hadron-nucleus and in nucleus-nucleus collisions at energies $\sqrt{s_{NN}}$

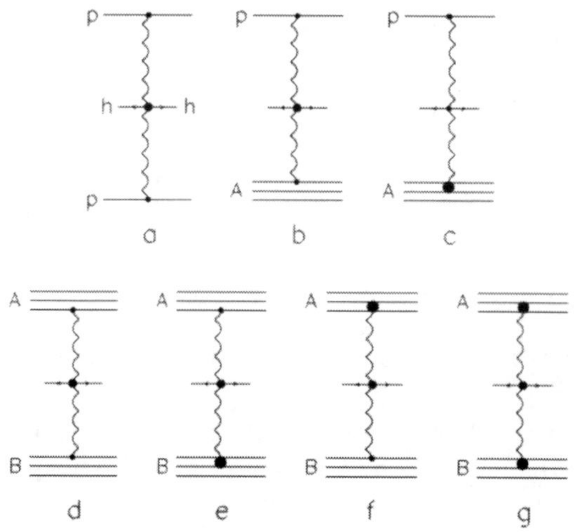

Figure 2. Reggeon diagrams for the different possibilities corresponding to the inclusive spectra of a secondary hadron h produced in the central region in (a) pp, (b,c) pA, and (d–g) AB collisions. Pomerons are shown by wavy lines.

= 14–30 GeV. At RHIC energies the situation drastically changes. The spectra of secondaries produced in pp collisions are described rather well [5], but the RHIC experimental data for Au+Au collisions [46, 47] give clear evidence of the inclusive density suppression effects which reduce by a factor ~0.5 the midrapidity inclusive density, when compared to the predictions based on the superposition picture [48, 49, 50]. This reduction can be explained by the inelastic screening corrections connected to multipomeron interactions [32] (see figure 3).

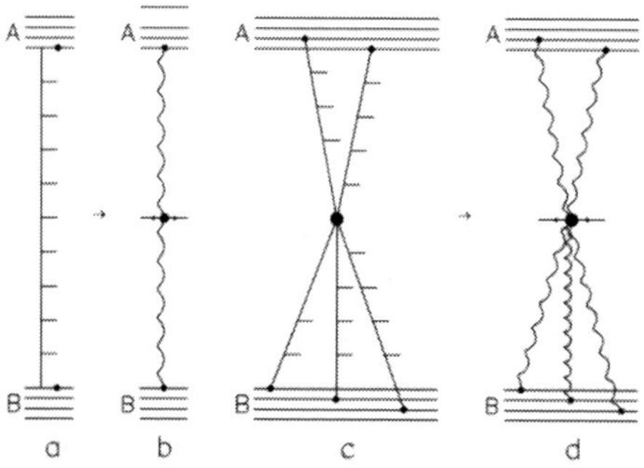

Figure 3. (a) Multiperipheral ladder corresponding to the inclusive cross section of diagram (b), and (c) fusion of several ladders corresponding to the inclusive cross section of diagram (d).

At fixed energies, $\sqrt{s_{NN}} \leq 30$–40 GeV, the nucleus-nucleus interaction can be described as the sum of one-Pomeron and of multipomeron eikonal exchanges. The inelastic processes are then determined by the production of one (figure 3a), or several (figure 3c) multiperipheral ladders, and the corresponding inclusive cross sections are described by the diagrams of figures 3b and 3d.

In accordance with the Parton Model [51, 52], the fusion of multiperipheral ladders shown in figure 3c becomes more and more important with the increase of the energy, resulting in the reduction of the inclusive density of secondaries. Such processes correspond to the enhancement Reggeon diagrams of the type of figure 3d, and to even more complicate ones. All these diagrams are proportional to the squared longitudinal form factors of both colliding nuclei [32]. Following the estimations presented in reference [32], the RHIC energies are just of the order of magnitude needed to observe this effect.

Unfortunately, all quantitative estimations are model dependent, since the numerical weight of the contribution of the multipomeron diagrams is rather unclear due to the many unknown vertices in these diagrams. However, the number of unknown parameters can be reduced in some models, and, as an example, in reference [32] the Schwimmer model [53] was used for the numerical calculations.

Another possibility to estimate the contribution of the diagrams with Pomeron interaction comes [54, 55, 56, 57, 58] from Percolation Theory. The percolation approach and its previous version, the String Fusion Model [59, 60, 61], predicted the multiplicity suppression seen at RHIC energies, long before any RHIC data were measured. New calculations of inclusive densities and multiplicities in percolation theory both in pp [62, 63], and in heavy ion collisions [63, 64], are in good agreement with the experimental data in a wide energy region.

In order to account for the percolation (inelastic screening) effects in the QGSM, it is technically more simple [23] to consider the maximal number of Pomerons n_{max} emitted by one nucleon in the central region that can be cut. These cut Pomerons lead to the different final states. Then the contributions of all diagrams with $n \leq n_{max}$ are accounted for as at lower energies. The larger number of Pomerons $n > n_{max}$ can also be emitted obeying the unitarity constraint, but due to the fusion in the final state (at the quark-gluon string stage), the cut of $n > n_{max}$ Pomerons results in the same final state as the cut of n_{max} Pomerons.

The QGSM fragmentation formalism allows one to calculate the integrated over p_T spectra of different secondaries as the functions of rapidity and x_F. The number of strings that can be used for the secondary production should increase with the initial energy [65]. The contribution coming from the coherent interaction of two nuclear clusters [31] has been estimated following [33] to be not larger than (20−30)%.

In this frame, we obtain a reasonable agreement with the experimental data on the inclusive spectra of secondaries produced in d+Au collisions at RHIC energy [23] with a value $n_{max} = 13$, and in p+Pb collisions at LHC energy [66] with the value $n_{max} = 23$.

However, in nucleus-nucleus collisions, even at not very high energies (e.g. see [35]), the yield of $\overline{\Xi}^+$ is more strongly enhanced than that of $\overline{\Lambda}$, and also the yield of $\Omega^- + \overline{\Omega}^+$ shows a stronger enhancement than the yield of $\overline{\Xi}^+$. This means that one additional source of multistrange hyperons must be at work from collective interactions of several strings.

At intermediate energies, we can think of rare situations when a few strings overlap in a small volume. This can be possibly considered as a QGP state effect. This can occur when the density of produced quark-gluon strings is several times larger than its average value.

When we consider central Pb+Pb collisions, we see that the experimental data on the ratios in equations 1 and 2 are in disagreement with the standard QGSM predictions. The experimental ratios show very strong energy dependences as a function of centrality in the CERN SpS-RHIC energy interval. To account for this effect, we can assume that some new contribution for multistrange antihyperon formation appears in this case. For example, the three strange antiquarks needed for the formation of $\overline{\Omega}^+$ could be taken at some energies out of three different quark-gluon strings, and not from the fragmentation of the same string. This could be possible if the multiplicity of newly produced strange-antistrange pairs would decreases with the number of pairs faster than, say, a Poissonian distribution. This new contribution is not included in the standard QGSM scheme, and it can be identified as a QGP formation signature.

Journal of Physics: Conference Series **607** (2015) 012006 doi:10.1088/1742-6596/607/1/012006

3. Numerical results

3.1. Net baryon spectra

We present the QGSM results for dn/dy of net proton $(p - \overline{p})$ and net Λ-hyperon $(\Lambda - \overline{\Lambda})$ productions at 200 GeV per nucleon both in proton-nucleus and in nucleus-nucleus collisions, and their comparison to the corresponding experimental data by the NA35 Collaboration [67]. The absolute normalization of dn/dy in all cases is determined by the data of proton and antiproton production in pp collisions at similar energies.

In the case of net proton production in $p-{}^{32}$S collisions, the QGSM calculations are in reasonable agreement with the experimental data. In the case of $p-{}^{197}$Au collisions the number of net protons is too small at small rapidities, what can be explained by the influence of the target fragmentation region. The nuclear cluster contribution, which is important mainly in the beam fragmentation region, seems to be too large.

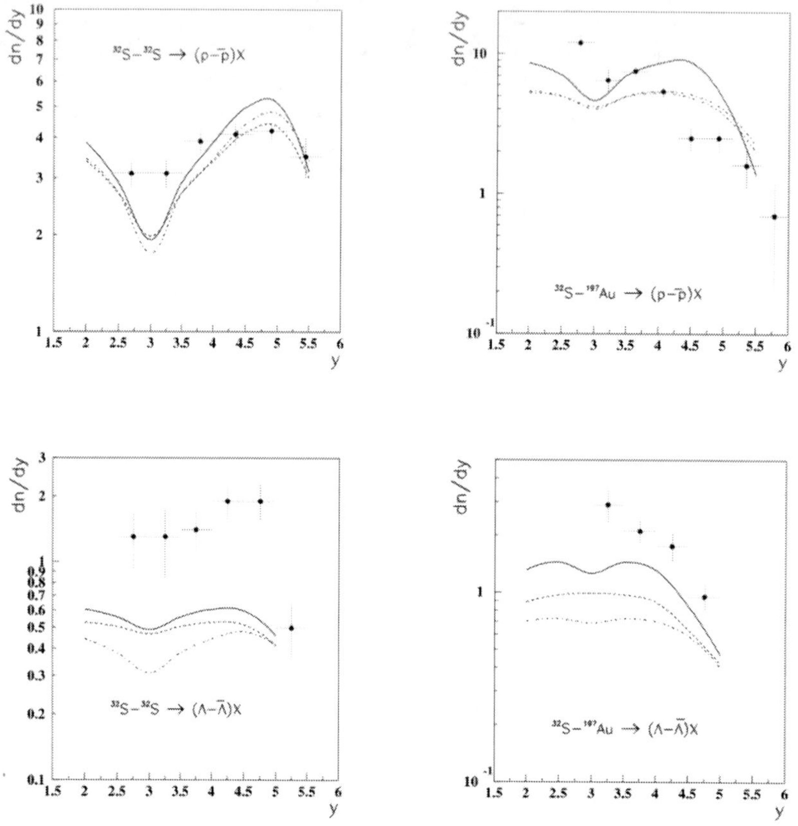

Figure 4. Net proton $p - \overline{p}$ (upper panels) and net Λ-hyperon $\Lambda - \overline{\Lambda}$ (lower panels) production in ^{32}S - ^{32}S (left panels) and in ^{32}S - ^{197}Au (right panels) collisions at 200 GeV per nucleon. Solid curves show the QGSM calculations with both SJ and cluster contributions, dashed curves with SJ contributions but without the cluster ones, and dotted curves without both SJ and cluster contributions.

In the case of net Λ-hyperon production the experimental error bars are rather large, and one can talk of general semiquantitative agreement of the QGSM calculations with the data.

For the sulphur ^{32}S beam shown in figure 4, the theoretical calculations are in reasonable agreement with the data for net proton production, whereas the data on net Λ-hyperon production are systematicaly higher than all calculated curves. In any case, though, all disagreements are not large than ~30%.

The data of the NA35 Collaboration [67] are compatible with the results of the NA44 Collaboration [68].

3.2. Production in midrapidity region

The NA49 Collaboration obtained experimental data [69, 70] for yields of Λ and $\overline{\Lambda}$ hyperons, and for $\overline{\Xi}^+$ in midrapidity region ($|y| < 0.4$ for Λ and $\overline{\Lambda}$, and $|y| < 0.5$ for $\overline{\Xi}^+$) in the central C+C, Si+Si, and Pb+Pb collisions (5% centrality for $\overline{\Lambda}$, and 10% centrality for $\overline{\Xi}^+$), at 158 GeV per nucleon.

A reasonable agreement with the QGSM predictions can be seen for secondary Λ and $\overline{\Lambda}$, but the calculated yields of $\overline{\Xi}^+$ are significantly smaller than those measured by the NA49 Collaboration.

On the other hand, the NA57 Collaboration obtained the experimental data [71] for Λ, $\overline{\Lambda}$, $\overline{\Xi}^+$, and $\overline{\Omega}^+$ yields in midrapidity region $|y| < 0.5$ in the minimum bias p+Be and p+Pb interactions, and in central (5% centrality) Pb+Pb collisions at 158 GeV per nucleon.

Unfortunately, the data by the NA49 and NA57 Collaborations are not compatible, the values of dn/dy for different hyperons measured by one collaboration being far outside the error bars of the corresponding values published by the other collaboration for the same centrality. This is probably due to different experimental event selection.

Here again one can see that the calculated yields of Λ and $\overline{\Lambda}$ are in agreement with experimental data on the level of 20−30% accuracy. Inclusive densities of $\overline{\Xi}^+$ hyperons are reasonably reproduced for the cases of p+Be and p+Pb collisions, but are several times underestimated in the case of central Pb+Pb interactions. For $\overline{\Omega}^+$ production in central Pb+Pb collisions, the disagreement is larger than one order of magnitude. The reason to discuss the production of antihyperons in these processes is that since a large baryon charge already exists in the initial state, some hyperons can be produced via the final state interaction of any baryon with a strange meson, but, on the contrary, the production of multistrange antihyperon production would be a signal of the production of several strange antiquarks in some small space-time volume.

At the LHC energy $\sqrt{s_{NN}} = 3$ TeV, the numerical QGSM predictions we obtain for dn/dy in the cases of midrapidity production of Λ and $\overline{\Lambda}$ in central Pb+Pb collisions are 36.2 and 35.6, respectively.

3.3. Ratios of multistrange to strange antihyperon production

These effects are more evidently seen in Fig. 5, where the experimental ratios of $\overline{\Omega}^+/\overline{\Lambda}$ and $\overline{\Xi}^+/\overline{\Lambda}$ as functions of the number of participating nucleons, N_w, together with the corresponding QGSM results, shown by solid curves. The first two left points in every panel correspond to p+Be and p+Pb collisions, and the other points correspond to Pb+Pb interactions with different centralities.

We consider the disagreement of the QGSM results with the experimantal values as a signal of the quantitatively large new contribution of $\overline{\Omega}^+$ and $\overline{\Xi}^+$ production to the inclusive cross section. This new contribution could appear when strange antiquarks are taken from different Pomerons (quark-gluon strings), in some collective interaction that can be seen as at the origin of Quark-Gluon Plasma formation.

Thus, the selection of events with $\overline{\Xi}^+$, and especially with $\overline{\Omega}^+$ production, would allow the definition of a sample enriched by Quark-Gluon Plasma formation events, and such a sample could be compared to a sample of events where only $\overline{\Lambda}$ hyperons are produced, to unravel QGP definite features.

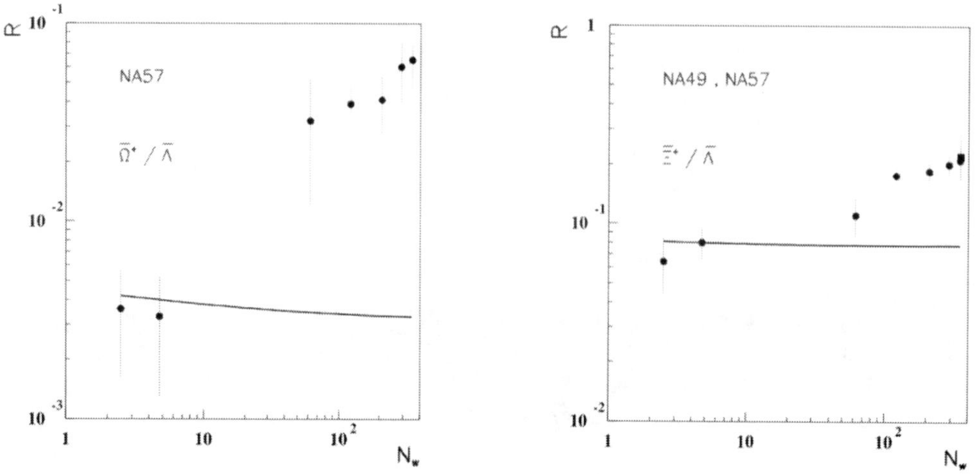

Figure 5. Ratios of $\overline{\Omega}^+$ to $\overline{\Lambda}$ (left panel), and of $\overline{\Xi}^+$ to $\overline{\Lambda}$ (right panel) as functions of the number of wounded nucleons, N_w. The experimental data for p+Be, p+Pb, and for Pb+Pb with different centralities measured by the NA57 Collaboration (points) and by the NA49 Collaboration (squares) are presented, together with the corresponding QGSM results, shown by solid curves.

Hyperon production at higher energies in midrapidity region was also measured at RHIC by the STAR Collaboration [72, 73] for Au + Au and Cu+Cu collisions at $\sqrt{s_{NN}} = 62.4$ GeV and 200 Gev.

The experimental ratios of $\overline{\Omega}^+$ to $\overline{\Lambda}$ obtained by the STAR Collaboration remain larger than the QGSM predictions at high centralities, and, so, the new mechanism by which the three strange antiquarks needed for $\overline{\Omega}^+$ formation are taken from different Pomerons is still important, though its relative contribution decreases in comparison with CERN SpS energies.

On the other hand, the experimental ratios of $\overline{\Xi}^+$ to $\overline{\Lambda}$ are now in agreement with the QGSM model calculations, probably meaning that the production of $\overline{\Xi}^+$ inside one only Pomeron becomes more effective at this energy than the possibility of taking two strange antiquarks from different Pomerons. This trend is confirmed by the STAR Collaboration data [73] at $\sqrt{s_{NN}}=$ 200 GeV.

Recently before this conference, the ALICE Collaboration has also published [74] results on the mid-rapiditiy anti-baryon to baryon ratios in *pp* collisions at $\sqrt{s}=$ 0.9, 2,76, and 7 TeV.

4. Conclusion

The QGSM provides a reasonable description of nucleon and Λ, as well as their antiparticles, production in nucleon-nucleus and nucleus-nucleus collisions at high energies. The level of numerical accuracy is of about 20−30%. Part of the uncertainty is connected to discrepancies among the different experimental data.

Inclusive densities of $\overline{\Xi^+}$ hyperons are reasonably reproduced for the cases of p+Be and p+Pb collisions [71], but they are several times underestimated in the case of central Pb+Pb interactions [71]. For $\overline{\Omega^+}$ production in central Pb+Pb collisions the disagreement is larger than one order of magnitude.

In nucleus-nucleus collisions, the number of quark-gluon strings (cut pomerons) increases with centrality. If the secondaries are independently produced in each quark-gluon string, the ratio of yields of different particles should not depend on centrality. However, it is experimentally clear (e.g. see [35], and figure 5) that the yield of $\overline{\Xi}^+$ is more strongly enhanced than that of $\overline{\Lambda}$, the same as the yield of $\Omega^- + \overline{\Omega}^+$ is more enhanced than that of $\overline{\Xi}^+$. This shows that an additional source of multistrange hyperons originated by the collective interactions of several strings must also play an actile role, and then this additional source of multistrange hyperons can possibly be considered as a QGP signature.

If this scenario in which collective effects among different quark-gluon strings help in explaining the energy dependence from CERN SpS to RHIC of the production ratios of $\overline{\Omega}^+$ to $\overline{\Lambda}$ and of $\overline{\Xi}^+$ to $\overline{\Lambda}$, would be confirmed, one could not expect a significant dependence of these ratios on centrality (see figure 5), neither significant differences of these ratios at LHC energies from the QGSM predictions, the expected values being $R(\overline{\Omega}^+/\overline{\Lambda}) \sim 0.14$ and $R(\overline{\Xi}^+/\overline{\Lambda}) \sim 0.014$.

A situation similar to the one shown in figures 5 can appear at LHC in Pb+Pb collisions in the midrapidity region for charmed antibaryon production, that is, for the ratios $\overline{\Omega}_c^+ = \overline{css}$ to $\overline{\Lambda}_c$, and $\overline{\Xi}_c^+$ to $\overline{\Lambda}_c$.

In summary, we can assume that the considered Quark-Gluon Plasma formation events are rather rare, but at the same time they can be clearly identified by a well-defined selection trigger. At CERN SpS energy, such a trigger can be the $\overline{\Omega}^+$ production in central Au+Au collisions, where a strong centrality dependence of the ratio $\overline{\Omega}^+$ to $\overline{\Lambda}$ is apparent (see figure 5a).

Acknowledgements

We want to thank Olga Piskunova for giving us the possibility to participate and contribute to this workshop. We are grateful to A.B. Kaidalov for useful discussions and comments. We also thank M. Poghosyan for his comments during this conference. This paper was supported by Ministerio de Economía y Competitividad of Spain (FPA2011−22776), the Spanish Consolider-Ingenio 2010 Programme CPAN (CSD2007-00042), by Xunta de Galicia, Spain (2011/PC043), by the State Committee of Science of the Republic of Armenia (Grant-13-1C023), and, partially, by grant RSGSS-3628.2008.2.

References

[1] A.B. Kaidalov and K.A. Ter-Martirosyan, *Yad. Fiz.* **39**, 1545 (1984); **40**, 211 (1984).
[2] A.B. Kaidalov and O.I. Piskounova, *Yad. Fiz.* **41**, 1278 (1985); *Z. Phys. C***30**, 145 (1986).
[3] Yu.M. Shabelski, *Yad. Fiz.* **44**, 186 (1986).
[4] G.H. Arakelyan, C. Merino, C. Pajares, and Yu.M. Shabelski, *Eur. Phys. J. C***54**, 577 (2008) and *hep-ph*/0709.3174.
[5] C. Merino, C. Pajares, and Yu.M. Shabelski, *Eur. Phys. J. C***71**, 1652 (2011).
[6] A.B. Kaidalov, K.A. Ter-Martirosyan, and Yu.M. Shabelski, *Yad. Fiz.* **43**, 1282 (1986).
[7] Yu.M. Shabelski, *Z. Phys. C***38**, 569 (1988).
[8] Yu.M. Shabelski, *Yad. Fiz.* **50**, 239 (1989).
[9] Yu.M. Shabelski, *Z. Phys. C***57**, 409 (1993).
[10] J. Dias de Deus and Yu.M. Shabelski, *Yad. Fiz.* **71**, 191 (2008).
[11] V.A. Abramovsky, V.N. Gribov, and O.V. Kancheli, *Yad. Fiz.* **18**, 595 (1973).
[12] A.B. Kaidalov, *Sov. J. Nucl. Phys.* **45**, 902 (1987); *Yad. Fiz.* **43**, 1282 (1986).
[13] Yu.M. Shabelski, *Yad. Fiz.* **26**, 1084 (1977); *Nucl. Phys. B***132**, 491 (1978).
[14] L. Bertocchi and D. Treleani, *J. Phys. G***3**, 147 (1977).
[15] J. Weis, *Acta Phys. Polonica B***7**, 85 (1977).
[16] T. Jaroszewicz *et al.*, *Z. Phys. C***1**, 181 (1979).
[17] V.M. Braun and Yu.M. Shabelski, *Int. J. Mod. Phys. A***3**, 2117 (1988).
[18] G.D. Alkhazov *et al.*, *Nucl. Phys. A***280**, 365 (1977).
[19] G.H. Arakelyan, A. Capella, A.B. Kaidalov, and Yu.M. Shabelski, *Eur. Phys. J. C***26**, 81 (2002) and *hep-ph*/0103337.
[20] F. Bopp and Yu.M. Shabelski, *Yad. Fiz.* **68**, 2155 (2005) and *hep-ph*/0406158; *Eur. Phys. J. A***28**, 237 (2006) and *hep-ph*/0603193.
[21] G.H. Arakelyan, C. Merino, and Yu.M. Shabelski, *Yad. Fiz.* **69**, 911 (2006) and hep-ph/0505100; *Phys. Atom. Nucl.* **70**, 1110 (2007) and *hep-ph*/0604103; *Eur. Phys. J. A***31**, 519 (2007) and *hep-ph*/0610264.

[22] O.I. Piskounova, *Phys. Atom. Nucl.* **70**, 1110 (2007) and *hep-ph*/0604157.
[23] C. Merino, C. Pajares, and Yu.M. Shabelski, *Eur. Phys. J. C***59**, 691 (2009) and *arXiv*:0802.2195[*hep-ph*].
[24] C. Merino, M.M. Ryzhinski, and Yu.M. Shabelski, *Eur. Phys. J. C***62**, 491 (2009); *Lecture given by Yu.M. Shabelski at The XLIII PNPI Winter School on Physics, St.Petersburg (Russia)*, February 2009, *arXiv*:0906.2659[*hep-ph*].
[25] X. Artru, *Nucl. Phys. B***85**, 442 (1975).
[26] M. Imachi, S. Otsuki, and F. Toyoda, *Prog. Theor. Phys.* **52**, 346 (1974); **54**, 280 (1976); **55**, 551 (1976).
[27] G.C. Rossi and G. Veneziano, *Nucl. Phys. B***123**, 507 (1977).
[28] D. Kharzeev, *Phys. Lett. B***378**, 238 (1996).
[29] L.L. Frankfurt and M.I. Strikman, *Phys. Rep.* **76**, 215 (1981).
[30] A.V. Efremov *et al.*, *Sov. J. Nucl. Phys.* **47**, 858 (1988).
[31] A.V. Efremov *et al.*, *Phys. Atom. Nucl.* **57**, 874 (1994).
[32] A. Capella, A. Kaidalov, and J. Tran Thanh Van, *Heavy Ion Phys.* **9**, 169 (1999).
[33] G.H. Arakelyan, C. Merino, and Yu.M. Shabelski, *arXiv*:1305.0388[*hep-ph*], and accepted for its publication in *Phys. Atom. Nucl.*
[34] G.H. Arakelyan, C. Merino, and Yu.M. Shabelski, *arXiv*:1402.6505[*hep-ph*].
[35] Huan Z. Huang, *J. Phys.* **G30**, 401 (2004).
[36] S.E. Vance, M. Gyulassy, and X-N. Wang, *Phys. Lett. B***443**, 45 (1998).
[37] A. Capella and B.Z. Kopeliovich, *Phys. Lett. B***381** 325 (1996).
[38] B.Z. Kopeliovich and B. Povh, *Z. Phys. C***75**, 693 (1997); *Phys. Lett. B***446**, 321 (1999); *Phys. Lett. B***469**, 9 (2000).
[39] V.V. Anisovich and V.M. Shekhter, *Nucl. Phys. B***55**, 455 (1973).
[40] A. Capella and C.A. Salgado, *Phys. Rev. C***60**, 054906 (1999).
[41] G.H. Arakelyan, A.B. Kaidalov, C. Merino, and Yu.M. Shabelski, *Phys. Atom. Nucl.* **74**, 126 (2011) and *arXiv*:1004.4074[*hep-ph*].
[42] K. Aamodt *et al.*, ALICE Collaboration, *Phys. Rev. Lett.* **6015**, 072002 (2010) and arXiv:1006.5432[*hep-ex*].
[43] D.I. Blokhintsev, *JETP* **13**, 1295 (1957).
[44] V.A. Abramovsky, O.V. Kancheli, and I.D. Mandzhavidze, *Sov. J. Nucl. Phys.* **13**, 630 (1971); *Yad. Fiz.* **13**, 1102 (1971).
[45] Yu.M. Shabelski, *Sov. J. Nucl. Phys.* **45**, 143 (1987); *Z. Phys. C***38**, 569 (1988).
[46] B.B. Back *et al.*, PHOBOS Collaboration, *Phys. Rev. Lett.* **85**, 3100 (2000).
[47] K. Adcox *et al.*, PHENIX Collaboration, *Phys. Rev. Lett.* **86**, 500 (2001).
[48] A. Capella, C. Merino, and J. Tran Thanh Van, *Phys. Lett. B***265** (1991) 415.
[49] Yu.M. Shabelski, *Z. Phys. C***57**, 409 (1993).
[50] N. Armesto and C. Pajares, *Int. J. Mod. Phys. A***15**, 2019 (2000).
[51] O.V. Kancheli, *JETP Lett.* **18**, 274 (1973).
[52] G.V. Davidenko and N.N. Nikolaev, *Yad. Fiz.* **24**, 772 (1976).
[53] A. Schwimmer, *Nucl. Phys. B***94**, 445 (1975).
[54] J. Dias de Deus, R. Ugoccioni, and A. Rodrigues, *Phys. Lett. B***458**, 402 (1999).
[55] J. Dias de Deus, R. Ugoccioni, and A. Rodrigues, *Eur. Phys. J. C***16**, 537 (2000).
[56] M.A. Braun and C. Pajares, *Phys. Rev. Lett.* **85**, 4864 (2000).
[57] J. Dias de Deus and Yu.M. Shabelski, *Eur. Phys. J. A***20**, 457 (2004).
[58] P. Brogueira, J. Dias de Deus, and C. Pajares, *Phys. Rev. C***75**, 054908 (2007).
[59] C. Merino, C. Pajares, and J. Ranft, *Phys. Lett. B***276**, 168 (1992).
[60] H.J. Möhring, J. Ranft, C. Merino, and C. Pajares, *Phys. Rev. D***47**, 4142 (1993).
[61] N.S. Amelin, M.A. Braun, and C. Pajares, *Z. Phys. C***63**, 507 (1994).
[62] I. Bautista, C. Pajares, and J. Dias de Deus, *Nucl. Phys. A***882**, 44 (2012).
[63] I. Bautista, J. Dias de Deus, G. Milhano, and C. Pajares, *Phys. Lett. B***715**, 230 (2012).
[64] I. Bautista, C. Pajares, G. Milhano, and J. Dias de Deus, *Phys. Rev. C***86**, 034909 (2012).
[65] J. Dias de Deus and C. Pajares, *Phys. Lett. B***695**, 211 (2012) and *arXiv*:1011.1099[*hep-ph*].
[66] C. Merino, C. Pajares, and Yu.M. Shabelski, *arXiv*:1207.6900[*hep-ph*].
[67] T. Alber *et al.*, NA35 Collaboration, *Eur. Phys. J. C***2**, 643 (2008) and *arXiv*:9711001[*hep-ex*].
[68] G. Bearden *et al.*, NA44 Collaboration, *Phys. Rev. C***57**, 837 (19988) and *arXiv*:9711004[*nucl-ex*].
[69] T. Anticic *et al.*, NA49 Collaboration, *Phys. Rev. C***80**, 034906 (2009) and *arXiv*:0906.0469[*nucl-ex*].
[70] T. Anticic *et al.*, NA44 Collaboration, *Phys. Rev.* **C78**, 034918 (2008) and *arXiv*:0804.3770[*nucl-ex*].
[71] F. Antinori *et al.*, NA57 Collaboration, *J. Phys. G***32**, 427 (2006) and *arXiv*:0601021[*nucl-ex*].
[72] M.M. Aggarwal *et al.*, STAR Collaboration, *Phys. Rev. C***83**, 024901 (2011) and *arXiv*:1010.0142[*nucl-ex*].
[73] G. Agakishiev *et al.*, STAR Collaboration, *Phys. Rev. Lett.* **108**, 072301 (2012) and *arXiv*:1107.2955[*nucl-ex*].
[74] E. Abbas *et al.*, ALICE Collaboration, *Eur. Phys. J. C***73**, 2496 (2013) and *arXiv*:1305.1562[*nucl-ex*].

Nonextensive thermodynamics with finite chemical potential, hadronic matter and protoneutron stars

Airton Deppman,[1] Eugenio Megías,[2,3] and Débora P. Menezes[4,5]

[1] *Instituto de Física, Universidade de São Paulo - Rua do Matão Travessa R Nr.187 CEP 05508-090 Cidade Universitaria, Sao Paulo - Brasil email: deppman@if.usp.br*

[2] *Grup de Fisica Teorica and IFAE, Departament de Fisica, Universitat Autonoma de Barcelona, Bellaterra E-08193 Barcelona, Spain*
[3] *Max-Planck-Institut fur Physik (Werner-Heisenberg-Institut), Fohringer Ring 6, D-80805, Munich, Germany email: emegias@mpp.mpg.de*

[4] *Departamento de Física, CFM, Universidade Federal de Santa Catarina, CP 476, CEP 88.040-900 Florianopolis - SC - Brazil*
[5] *Departamento de Fisica Aplicada, Universidad de Alicante, Ap. Correus 99, E-03080, Alicante, Spain email: debora.p.m@ufsc.br*

The nonextensive thermodynamics of an ideal gas composed by bosons and/or fermions is derived from its partition function for systems with finite chemical potentials. It is shown that the thermodynamical quantities derived in the present work are in agreement with those obtained in previous works when $\mu \leq m$. However some inconsistencies of previous references are corrected when $\mu > m$. A discontinuity in the first derivatives of the partition function and its effects are discussed in details. We show that at similar conditions, the nonextensive statistics provide a harder EOS than that provided by the Boltzmann-Gibbs statistics. This fact induced us to consider an application to a proto-neutron star, which is discussed for different assumptions.

PACS numbers: 05.70.Ce,95.30.Tg,26.60.-c

I. INTRODUCTION

One of the main objectives in the study of ultrarelativistic collisions is the investigation of the properties of the quark-gluon plasma. Thermodynamical aspects of the plasma are of specially interest due to the possibility of studying the deconfinement process. The nonextensive statistics proposed by C. Tsallis in 1988 [1] was introduced in High Energy Physics (HEP) by Bediaga, Curado and Miranda (BCM) [2] where the transverse momentum (p_T) distribution from Hagedorn's theory [3] was formally modified by the introduction of the q-exponential function commonly used in the Tsallis statistics. The resulting formula describes the whole p_T-distribution obtained in HEP experiments.

The nonextensive formula for p_T-distribution has been used in a large number of works, all of them reporting good agreement with experiment [4–8]. In Ref. [10] it has been shown that Hagedorn's theory can be generalized to include the nonextensive statistics [9, 11–15] self-consistently. The resulting theory shows that there must exist a limiting temperature, T_o, and also an entropic index, q_o, which characterizes the hadronic systems. The theory gives a new formula for the hadronic mass spectrum in terms of T_o and q_o which can describe quite well the known spectrum of hadrons with masses from the pion mass up to ~ 2.5 GeV [16]. In this way the nonextensive self-consistent theory [10] presents much more restrictive conditions to the applicability of the Tsallis statistics in HEP. Several analyses of experimental data [16–19] have shown that the theoretical predictions are in agreement with the experimental findings.

A study of the nonextensive thermodynamics of hadronic matter for null chemical potential (μ) and its comparison to Lattice-QCD data was already presented in Ref. [20] showing good agreement between LQCD and the thermodynamical theory. In this work we discuss the extension of the thermodynamics theory to finite chemical potential systems, which is of importance in the study of nucleus-nucleus collisions and of astrophysical massive objects.

The paper is organized as follows: in section 2 we introduce the partition function and show that it is in agreement with the p_T-distribution used in previous works [2, 10, 19, 21]; in section 3 we derive the thermodynamical functions of interest and establish the phase transition line between confined and deconfined regimes in the $T \times \mu$ diagram; in section 4 we discuss our results and in section 5 we present our conclusions.

The paper is organized as follows: in section II we introduce the partition function and show that it is in agreement with the p_T-distribution used in previous works [2, 10, 19, 21]; in section III we derive the thermodynamical functions of interest and establish the phase transition line between confined and deconfined regimes in the $T \times \mu$ diagram; in section IV we apply our results to the neutron star problem and finally we present our conclusions in section V. A full description of the thermodynamical theory developed here can be found in Ref. [22], and an application of the theory for (proto)neutron stars which is based on constant entropy can be found in Ref. [23] (see also Ref. [24]).

II. PARTITION FUNCTION FOR HADRONIC MATTER

The q-exponential function is defined as

$$\begin{cases} e_q^{(+)}(x) = [1+(q-1)x]^{(1/(q-1)} & , x \geq 0 \\ e_q^{(-)}(x) = \frac{1}{e_q^{(+)}(|x|)} = [1+(1-q)x]^{1/(1-q)} & , x < 0 \,. \end{cases} \tag{1}$$

We define the partition function for a nonextensive ideal quantum gas as

$$\ln_q \Xi(V,T,\mu) = \begin{cases} \frac{V}{(2\pi)^3} \int_m^\infty \log_q^{(+)} \left(\frac{e_q^{(+)}(x)+\xi}{e_q^{(+)}(x)} \right) \\ \frac{V}{(2\pi)^3} \int_m^\infty \log_q^{(-)} \left(\frac{e_q^{(-)}(x)+1}{e_q^{(-)}(x)} \right) \end{cases} , \tag{2}$$

where $x = \beta(\sqrt{p^2+m^2} - \mu)$, m is the hadron mass, $\beta = 1/T$ is the inverse of temperature, and μ the chemical potential with $\xi = 1$ for fermions and $\xi = -1$ for bosons. In the limit $q \to 1$ the partition functions defined above reduces to the well-known Fermi-Dirac and Bose-Einstein partition functions for fermions and bosons, respectively. Below we show that Equations 2 define in fact the partition function for the ideal quantum gas in Tsallis statistics.

III. THERMODYNAMICAL FUNCTIONS

To show that the partition function defined in Eq. (2) corresponds to the one for a quantum gas in non extensive statistics we derive the occupation number, average number of particles, energy density and the entropy, and show that the results are identical to the ones obtained in Refs [19, 21] in the sector $x \geq 0$. However, some differences are found and discussed for $x < 0$.

A. Derivation of thermodynamic quantities from the partition function

In this section we will derive from the partition functions defined in Eq. (2) the occupation number and the entropy and show that the results are identical to the results obtained in Refs [19, 21] for quantum gases in nonextensive statistics.

The occupation number can be obtained through the relations

$$n(\varepsilon) = \beta^{-1}\frac{\partial}{\partial\mu}\ln_q \Xi, \tag{3}$$

resulting in

$$n_q^{(+,-)}(p) = \left(\frac{1}{e_q^{(+,-)}(x) - \xi}\right)^{\tilde{q}}. \tag{4}$$

This result is identical to those obtained by CMP [21, 25] and by Cleymans and Worku [19]. In addition, observe that for $\mu = 0$ and ε sufficiently high it reduces to the equation used by BCM in Ref. [2] to describe the HEP p_T-distributions. Thus we can conclude that the occupation number derived from the partition functions defined in Eq. (2) can correctly describe the results obtained in ultrarelativistic collisions.

The entropy can be obtained through the relations

$$S = \beta^2\frac{\partial}{\partial\beta}\left(\frac{\ln_q \Xi_q}{\beta}\right), \tag{5}$$

resulting

$$S = V\int\frac{d^3p}{(2\pi)^3}\sum_{r=\pm}\Theta(rx)\Big[-[\bar{n}_q^{(r)}(x)]^{\tilde{q}}\log_q^{(-r)}\left(\bar{n}_q^{(r)}(x)\right) + \xi[1+\xi\bar{n}_q^{(r)}(x)]^{\tilde{q}}\log_q^{(-r)}\left(1+\xi\bar{n}_q^{(r)}(x)\right)\Big], \tag{6}$$

where we have defined $\bar{n}_q^{(r)}(x) \equiv [n_q^{(r)}(x)]^{1/\tilde{q}}$, which is identical to the CMP entropy defined in Ref [21].

With this results we conclude that the partition function in Eq. (2) does represent the relevant function for an ideal quantum gas in Tsallis statistics. Observe that for $\mu = 0$ and ε sufficiently high the partition functions defined here are similar to those used in Ref. [20] to extend Hagedorn's theory to nonextensive statistics. Therefore definitions given here are in accordance with the nonextensive self-consistent thermodynamics. In the following we explore some of the features of the thermodynamical systems described by Eq. (2).

B. Thermodynamical properties of hadronic systems

Before studying the properties of a thermodynamically equilibrated hadronic system in the non extensive thermodynamics we have to find the region where this system can exist. In fact, due to the transition from confined to deconfined regimes, the hadronic matter can be found only below the phase transition line. To find this line we use a result obtained by Cleymans and Redlich [26] where they pointed out that the transition line can be determined by the condition that $\langle E\rangle/\langle N\rangle = 1$ GeV. This result was obtained through a systematic analysis of particle yields from HEP experiments, where the authors used the Boltzmann's statistics in the formulation of their thermodynamics. But since yields are calculated by the integration over all energy states, and considering that the differences between the distributions obtained through Boltzmann statistics and Tsallis statistics are relevant only in the high energy tail, we can assume with some confidence that the same relation holds in the non extensive case. Of course this assumption must be checked by a similar analysis of experimental data, now using the non extensive formulas derived here, a task that is beyond the scope of the present work. However we present below some evidences that this hypothesis is correct.

In the following we assume that the entropic index, q_o, is a fixed property of the hadronic matter with its value determined in the analysis of p_T-distributions and in the study of the hadronic mass spectrum in Ref. [16], so we set $q_o = 1.14$, although in some cases we analyze the behavior of some quantities for different values of q. In the following we refer to Boltzmann-Gibbs statistics also by $q = 1$.

The system of interest here is a gas composed by different hadronic species in thermodynamical and chemical equilibrium. The partition function is then given by

$$\log\Xi_q(V,T,\{\mu\}) = \sum_i \log\Xi_q(V,T,\mu_i), \tag{7}$$

Mesons	Mass	S	g	Mesons	Mass	S	g
π^0	134.98	0	1	K^+	493.68	1	1
π^+, π^-	139.57	0	2	K^0	497.67	1	1
η	547.3	0	1	$K^*(892)^+$	891.66	1	3
$\rho(770)$	771.1	0	9	$K^*(892)^0$	896.1	1	3
$\omega(782)$	782.57	0	3	$K_1(1270)$	1273	1	6
$\eta'(958)$	957.78	0	1	$K_1(1400)$	1402	1	6
$f_0(980)$	980	0	1	$K_0^*(1430)$	1412	1	2
$a_0(980)$	984.7	0	3	$K^*(1410)$	1414	1	6
$\phi(1020)$	1019.46	0	3	$K_2^*(1430)^+$	1425.6	1	5
$h_1(1170)$	1170	0	3	$K_2^*(1430)^0$	1432.4	1	5
$b_1(1235)$	1229.5	0	9	$K(1460)$	1460	1	10
...	...	...	...	...	...	...	...

Table I. List of the lowest-lying mesons used in Eq. (7). We include some of their properties: mass, strangeness (S) and degeneracy (g). The corresponding anti-mesons with $S = -1$ are not shown in the table but they are considered as well in the computation. The dots indicate that heavier mesons are included in the computation, although they are not explicitly shown in this table due to lack of space.

where μ_i refers to the chemical potential for the *i-th* hadron. The lowest-lying hadrons considered in our calculations are taken from the Particle Data Group [27], and some of them are presented in Tables I and II. The total numbers of hadronic states considered are 808 for mesons and 1168 for baryons (+ anti-baryons), including degeneracies, which correspond to a maximum value of the mass about 11 GeV and 5.8 GeV respectively. The computation will be performed by restricting the ensemble summation in Eq. (7) to zero strangeness (see [26] for details). We will focus first on a study of the phase transition by assuming $\mu = 0$ for all mesons and considering that all baryons have the same chemical potential value $\mu_B \neq 0$. After that, the effect of a nonzero value of the chemical potential for pions will be explored as well.

The phase transition line obtained as described above is reported in Fig. 1 (left), and it is compared to the Boltzmann's result, corresponding to $q \to 1$. For the sake of clarity we refer to the temperature obtained with BG statistics as T and to the temperature obtained with Tsallis statistics as τ. The relation between τ and T was already investigated in Ref. [20], and it was found that for a fixed chemical potential there is a linear correspondence between both quantities. The relation between both quantities when varying μ_B is shown in Fig. 1 (right).

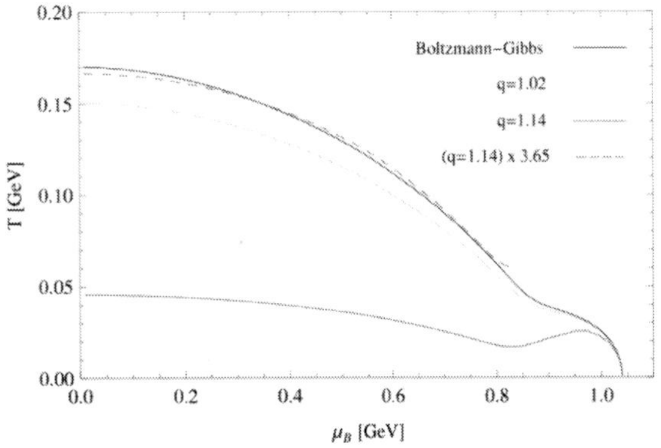

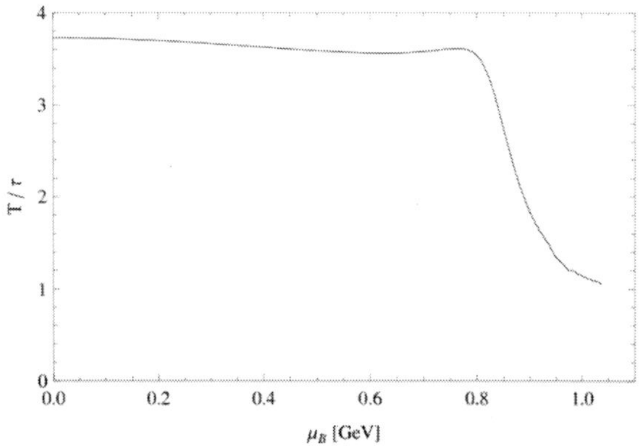

Figure 1. Left: Chemical freeze-out line $T = T(\mu_B)$ obtained by assuming $\langle E\rangle/\langle N\rangle = 1$ GeV. We plot as continuous lines from top to bottom the result using: Boltzmann-Gibbs statistics, Tsallis statistics with $q = 1.02$, and with $q = 1.14$. Dashed red line corresponds to the curve with $q = 1.14$ multiplied by the factor 3.65. Right: Quotient between physical temperature T obtained with BG statistics, and the effective temperature τ obtained with Tsallis statistics and $q = 1.14$, as a function of the baryonic chemical potential. We consider $\mu = 0$ for mesons in both figures.

Baryons	Mass	S	g	Baryons	Mass	S	g
p	938.27	0	2	Λ^0	1115.68	-1	2
n	939.56	0	2	Σ^+	1189.37	-1	2
$\Delta(1232)$	1232	0	16	Σ^0	1192.64	-1	2
$N(1440)$	1440	0	4	Σ^-	1197.45	-1	2
$N(1520)$	1520	0	8	Σ^{*+}	1382.8	-1	4
$N(1535)$	1535	0	4	Σ^{*0}	1383.7	-1	4
$\Delta(1600)$	1600	0	16	Σ^{*-}	1387.2	-1	4
$\Delta(1620)$	1620	0	8	$\Lambda(1405)$	1406	-1	2
...	...	...	...	...	...	...	...
Ξ^0	1314.83	-2	2	Ω^-	1672.45	-3	4
Ξ^-	1321.31	-2	2				
Ξ^{*0}	1531.8	-2	4				
Ξ^{*-}	1535	-2	4				
...	...	...	...				

Table II. List of the lowest-lying baryons used in Eq. (7). We show only the baryons with baryonic number $B = 1$. The corresponding anti-baryons $B = -1$ and $S = 0, 1, 2$ and 3 are considered as well in the computation.

We observe in Fig. 1 (left) that for both $q = 1$ and $q = 1.14$ we obtain lines with similar shapes, but τ is always lower than T. The ratio between both temperatures is practically constant up to $\mu_B \simeq 0.8$ GeV, as can be seen in Fig. 1 (right), and a consequence is that the traced line which results from the multiplication of the results with $q = 1.14$ by a constant factor equal to 3.65 almost reproduces the result with BG statistics in Fig. 1 (left). We also include in Fig. 1 (left) the results for an intermediate value of $q = 1.02$, which shows that the phase transition line smoothly approaches that obtained for Boltzmann statistics as q decreases towards the unity. All curves in Fig. 1 (left) show an inflexion for $\mu_B \sim 0.9$ GeV which is related to the sharp increase in the baryon density as the baryonic chemical potential approaches the proton/neutron mass $m_{p,n} \simeq 0.94$ GeV. This increase is displayed in Fig. 2, where we plot the proton and neutron densities as a function of μ_B in Tsallis statistics. A similar behaviour is observed in Boltzmann-Gibbs statistics.

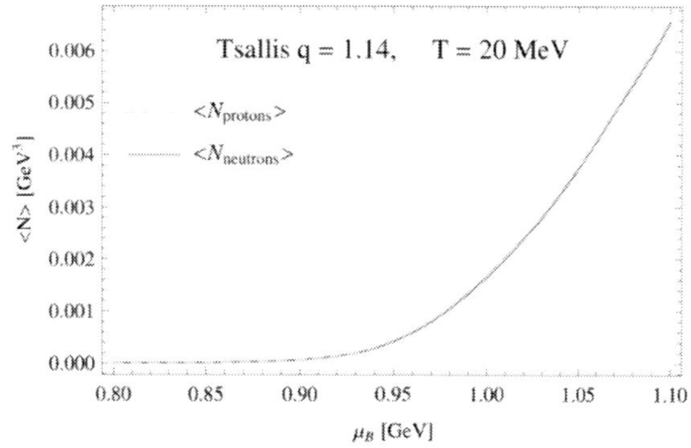

Figure 2. Density of protons and neutrons as a function of the baryonic chemical potential, in Tsallis statistics with $q = 1.14$. We consider $T = 20$ MeV. Dashed green line corresponds to the density of protons, while continuous red line is the density of neutrons.

For $\mu_B = 0$ the effective temperature is $T_o = 45.6$ MeV for $q = 1.14$, which is not in agreement with the value $T_o = (60.7 \pm 0.5)$ MeV found in the analysis of the p_T-distributions [16]. This disagreement can be related to the value adopted for $\langle E \rangle / \langle N \rangle$, which still must be checked by analysis of experimental data with the non extensive statistic. In order to provide an estimate of the sensitivity of the value of T_o when changing q it is worth mentioning that the effective temperature for a slightly smaller value of the entropic index, $q = 1.12$, is $T_o = 61.0$ MeV, which is in

Journal of Physics: Conference Series **607** (2015) 012007 doi:10.1088/1742-6596/607/1/012007

agreement with that reference.

The transition line determines the region where the confined states exist (below the line), and the region where one expects to find the quark-gluon plasma (above the line).

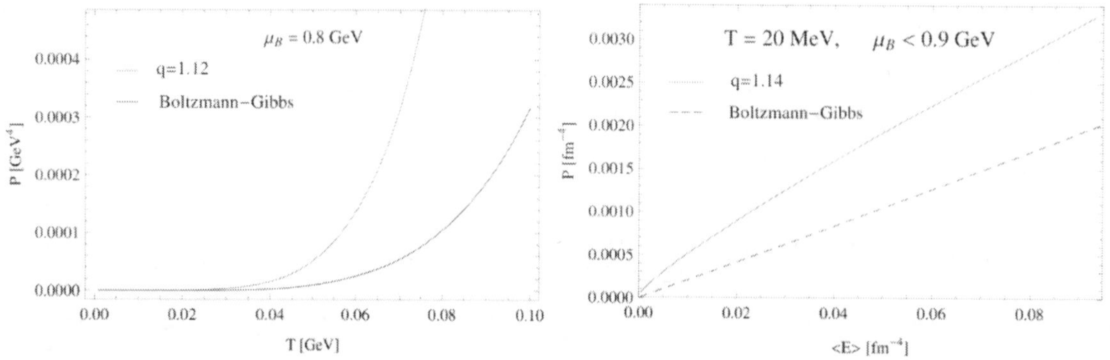

Figure 3. Left: Pressure as a function of temperature. The baryonic chemical potential is kept fixed to the value $\mu_B = 0.8$ GeV. Right: Pressure as a function of energy density when changing the baryonic chemical potential in the range 0 GeV $< \mu_B <$ 0.9 GeV. The temperature is kept fixed to the value $T = 20$ MeV. In both figures we plot as a dashed blue line the result in Boltzmann-Gibbs statistics, and as a continuous red line the result in Tsallis statistics with $q = 1.14$.

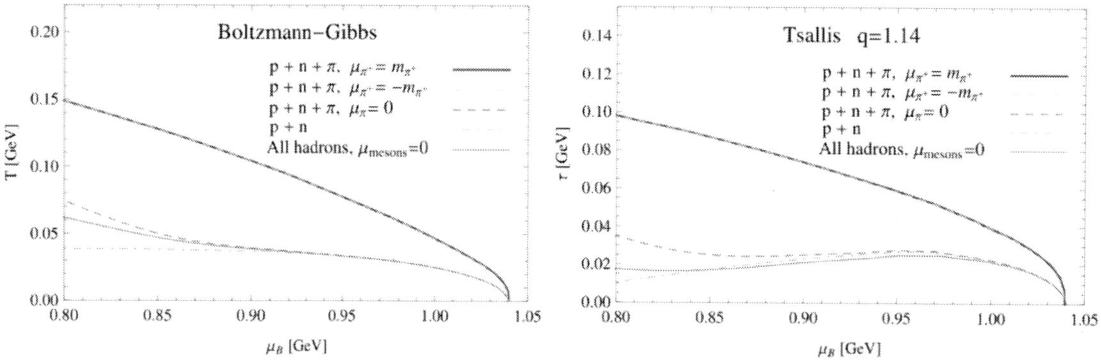

Figure 4. Chemical freeze-out line $T = T(\mu_B)$ obtained by assuming $\langle E\rangle/\langle N\rangle = 1$ GeV. Left: Boltzmann-Gibbs statistics. Right: Tsallis statistics with $q = 1.14$. We show the result in different cases, including: i) "(anti)protons + (anti)neutrons", ii) "(anti)protons + (anti)neutrons + pions" with $\mu_\pi = 0$, $\mu_{\pi^+} = m_{\pi^+}$ or $\mu_{\pi^+} = -m_{\pi^+}$, iii) all hadrons with $\mu_{\rm mesons} = 0$ (as in Fig. 1). In ii) we consider $\mu_n = \mu_p - \mu_{\pi^+}$, see Eq. (13), and in this case it is represented in the horizontal axis μ_p when $\mu_{\pi^+} = m_{\pi^+}$, and μ_n when $\mu_{\pi^+} = -m_{\pi^+}$.

In Fig. 3 (left) we plot the pressure as a function of the temperature for the chemical potentials fixed at the value $\mu_B = 0.8$ GeV for the baryonic chemical potential and $\mu = 0$ for all mesons with $q = 1$ and $q = 1.14$. We observe that the pressure increases faster in the non extensive case, $q = 1.14$, than in the extensive one, $q = 1$. In Fig. 3 (right) we show the results for the pressure as a function of the energy density for $T = 20$ MeV. Also here we observe that the pressure increases faster in the case of $q = 1.14$ in comparison with the case $q = 1$. From Fig. 3 one can conclude that the EOS for hadronic matter obtained with the Tsallis statistics is harder than the one obtained from the BG case.

Up to now we have studied the equation of state including the spectrum of hadrons in Table I and II. It would be interesting to analyze also the case in which only protons, neutrons and possibly pions contribute to the equation of state, all of them with nonzero chemical potential. We have studied the finite pion chemical potential in two different cases: a) $\mu_{\pi^+} = m_{\pi^+} = -\mu_{\pi^-}$, b) $\mu_{\pi^+} = -m_{\pi^+} = -\mu_{\pi^-}$; and $\mu_{\pi^0} = 0$, $\mu_n = \mu_p - \mu_{\pi^+}$ in both cases. These values are motivated in section IV. The phase transition lines in Boltzmann-Gibbs and in Tsallis statistics in the regime of high baryonic chemical potential are plotted in Fig. 4. Note that the effect of finite pion chemical potential is to increase the transition line to higher values of temperature, and the curves with $\mu_{\pi^+} = m_{\pi^+}$ and $\mu_{\pi^+} = -m_{\pi^+}$ coincide, as we have not introduced electrons in the computation (see section IV for a discussion including electrons). We show in Fig. 5 the result for the equation of state considering zero and finite pion chemical potential. The effect of pions is to increase the values for the pressure with respect to the case with only protons and neutrons. When considering a

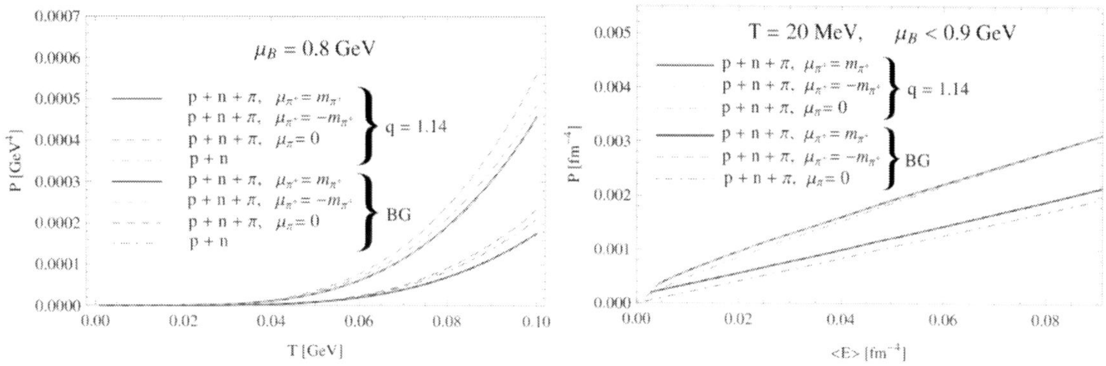

Figure 5. The same as in Fig. 3, but including only the lowest states in the hadron spectrum, i.e. "(anti)protons + (anti)neutrons" and "(anti)protons + (anti)neutrons + pions". In the latter case, we plot curves with either $\mu_\pi = 0$, $\mu_{\pi^+} = m_{\pi^+}$ and $\mu_{\pi^+} = -m_{\pi^+}$. For the clarity of the figure it is not plotted the result for p + n in the right figure, as it is very close to the result with p + n + π, $\mu_\pi = 0$. We display the results using Boltzmann-Gibbs (lower curves) and Tsallis statistics with $q = 1.14$ (higher curves). See Figs. 3 and 4 for other details.

nonzero value for the pion chemical potential this leads to a noticeable effect on the EOS, as it becomes harder either in Boltzmann-Gibbs or in Tsallis statistics, see Fig. 5 (right).

The results presented in this section make us to consider an application to stellar matter, where high pressures are necessary to compensate for the gravitational force so that protoneutron star stability and correct macroscopic properties are attained.

IV. APPLICATION TO (PROTO)NEUTRON STARS

During the evolution process of a protoneutron star, neutrinos are believed to be trapped for a few seconds and then leave the star carrying huge amounts of energy. This process is known as deleptonization. The entropy then increases and the cooling process starts until the final temperature is assumed to be zero because the thermal energy is lower than the Fermi energy of the system [28, 29]. During these processes, simulations have shown that the stellar temperature can reach values of the order of 60 MeV in a nucleonic star or lower, if the star is a hybrid object with a quark core [30]. Of course, the temperature inside a star is not expected to be constant. According to simulations of protoneutron star cooling, the entropy per baryon (s) can reach up to 2 in Boltzmann units. Hence, the ideal situation would be to consider fixed entropy scenarios, but for academic purposes and until we have a better understanding of the physics involved in non extensive thermodynamics, we shall start with fixed temperatures of the order of $\tau = 20$ MeV. The calculation of equations of state at fixed temperature already gives a reasonable understanding of the overall behaviour of protoneutron star matter [30, 31]. In the present work, the main idea is to compare the results obtained with non-extensive statistics with the ones obtained with the usual Fermi gas and the inclusion of entropy and trapped neutrinos (necessary only for the $s = 1$ case), do not modify the general picture.

When one wants to apply the results presented in the last sections to the study of neutron stars, some important ingredients also have to enter in the calculations. As explained in the Introduction, besides the hadronic matter, leptons are also present in stellar matter and these particles play a relevant role in the equilibration between gravitational force and the gas pressure due to their small masses. One also needs to make sure that the neutron stars are electrically neutral.

For the sake of simplicity let us initially think about a neutron star composed only by protons, neutrons, electrons and muons. The star is not static, it is indeed a dynamically equilibrated system because protons and neutrons are being converted into one another through the weak decays:

$$\begin{cases} n \to p + e^- + \bar{\nu}_e \\ p \to n + e^+ + \nu_e \end{cases} \tag{8}$$

and the dynamical equilibrium between the relative number of neutrons and protons is regulated by their respective chemical potentials, which must satisfy the relation

$$\mu_n = \mu_p + \mu_e \quad \mu_e = \mu_\mu , \tag{9}$$

where it was assumed that the positron chemical potential is opposite to the electron chemical potential. The equations above cannot fully determine μ_n, so this is a free parameter which is in fact the one which regulates the baryonic density in the star, given by

$$\rho = \rho_p + \rho_n \,, \tag{10}$$

where ρ_p and ρ_n are respectively the proton and neutron densities. As the star has to be electrically neutral, the condition that

$$\rho_p = \rho_e + \rho_\mu \tag{11}$$

is also enforced. Therefore our first calculation is for a neutron-proton-lepton star with leptons being included in our EOS by using the standard BG statistics for a fermionic free gas. Notice that we have two independent chemical potentials, μ_n and μ_e, which are constrained by Eqs. (9) and (11). It is also well known that the macroscopic star properties cannot be explained in terms of free Fermi gases only. In order to attain a reasonable stellar mass, all models have to take into account the nuclear interaction, generally described by mediating mesons of Yukawa type interactions [32]. In [33], protoneutron stars with a fixed entropy per baryon $s = 2$ were studied within quantum hadrodynamics extended to consider also nonextensive statistical effects. The difference between the usual non-linear Walecka type models [30, 32, 34] and the work developed in [33] is only due to the different fermion distribution functions for hadrons (the paper does not clarify whether the Tsallis statistics is used for leptons as well). Bearing in mind that, apart from the fact that the equations of motion in a mean field approximation are solved in a self-consistent way, the influence of the nuclear interaction, mediated by the meson fields, is approximately the same in both approaches. Hence, the main influence of the non-extensive statistics resides in the gas contribution, which we analyse next.

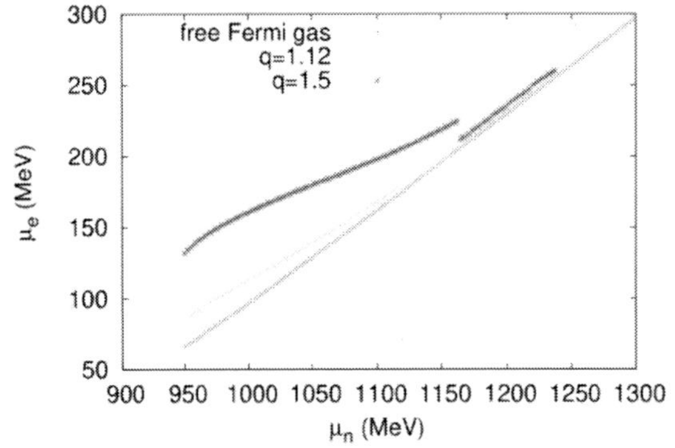

Figure 6. Electron chemical potential as a function of the neutron chemical potential for BG and Tsallis statistics.

In Fig. 6 we show how the electron chemical potential varies with the neutron chemical potential in different situations: 1) neutron, proton and leptons described by free Fermi gases, 2) neutrons and protons described by nonextensive thermodynamics with $q = 1.14$, following the analysis performed in Ref. [16], and leptons described by a free Fermi gas, and 3) neutrons and protons described by non extensive thermodynamics with $q = 1.5$ and leptons described by a free Fermi gas. One can see that electron chemical potential increases with the increase of the q value, resulting in different amounts of particles inside the star, which are depicted in Fig. 7. One can observe that there is a discontinuity, which appears at larger neutron chemical potentials as q increases and that reflects itself in the particle fractions. After the discontinuity, the resulting electron chemical potential tends to the results obtained with the BG statistics. The particle density integrals run from zero to infinity and it can be divided into two integrals, one running from zero to $\sqrt{\mu_b^2 - m_b^2}$, where $b = p, n$ and the other running from this point to infinity. The discontinuities take place exactly at the point where the momentum reaches the values $\sqrt{\mu_b^2 - m_b^2}$ and it coincides for protons and neutrons. The effects of these discontinuities have already been discussed in the last section. We define the particle fraction as $Y_i = \rho_i/\rho$, where ρ is the baryonic density. From Fig, 7, we can see that the particle fractions change considerably from one case to the other, when we move from the BG statistics to the non-extensive statistics.

In the simplest mean-field approximation to neutron star matter using the non-linear Walecka model [32], the pion contribution vanishes and, therefore, they are not considered as constituent particles of the star. In more sophisticated methods, as the Hartree-Fock approximation, they can be introduced. In our approach, it is easy to include pions,

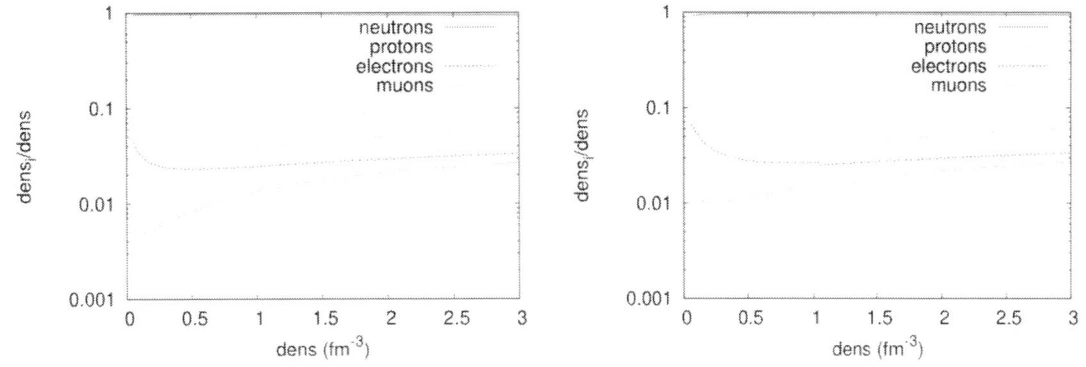

Figure 7. Particle fractions as a function of the baryonic density at $\tau = 20$ MeV for: left) free Fermi gas, right) Tsallis statistics with $q = 1.14$.

since they are automatically considered in the thermodynamics at high energies, so we have decided to study also the role of pions in the neutron star stability in our second calculation where the star is composed by neutrons, protons, pions and leptons. Pions usually decay into muons through the weak interaction, but in a dense hadronic medium as a neutron star it is more probable that they interact with protons and neutrons through the strong interaction. Therefore, we next check whether the pions can be partially responsible for the nuclear forces acting in the star, in such a way that the number of pions is regulated by their chemical potential and in turn, the pions regulate the intensity of the nuclear force. It is important to emphasize that we do not expect the pions to completely mimic the role of the nuclear interaction in this simple picture, since we know, from the previous sections, that the non-extensive thermodynamics acts as a small *perturbation or correction* in the system.

The pionic chemical potential can vary inside the limit $\mu_\pi < m_\pi$ and for positive and negative pions the relation $\mu_{\pi^+} = -\mu_{\pi^-}$ must be satisfied since one is the antiparticle of the other. For the same reason the neutral pions have null chemical potential, $\mu_{\pi^o} = 0$. In the nuclear interaction pions can convert protons into neutrons and vice-versa, through the processes

$$\begin{cases} n + \pi^+ \to p, \\ p + \pi^- \to n, \end{cases} \tag{12}$$

which lead to the relation

$$\mu_n = \mu_p - \mu_{\pi^+} \tag{13}$$

and the total electric charge has to be kept neutral so that

$$\rho_p + \rho_{\pi^+} = \rho_{\pi^-} + \rho_e + \rho_\mu \tag{14}$$

with

$$-m_{\pi^+} \leq \mu_{\pi^+} \leq m_{\pi^+}. \tag{15}$$

Once again, the leptons are included in our EOS by using the standard BG statistics for a fermionic free gas, and then we can calculate all the thermodynamical quantities of interest.

This is a quite simple and naive model, and can be improved in several ways, but it will be enough for a study of the role played by non extensive thermodynamics in neutron stars. In particular, despite the fact that the pions appear as intermediate particles for the nuclear force, we consider below just their on-shell mass and we have taken the pion mass equal to 135 MeV for the three species. Also other hadrons, as baryonic resonances heavier than proton and neutron could be present in the star but they are not included in our model.

In Fig. 5, we can see that the EOS obtained with a pion chemical potential of the order of the pion mass is the hardest of all. For this reason, we have considered three different possibilities for the pion chemical potential: i) the first one is to assume that $\mu_{\pi^-} = \mu_e$ or analogously that $\mu_{\pi^+} = -\mu_e$, what can be inferred directly from the comparison of Eqs. (9) and (13). The second possibility is to fix the pion chemical potential as the pion mass and we have taken the two limits of Eq. (15): ii) $\mu_{\pi^-} = m_\pi$ ($\mu_{\pi^+} = -m_\pi$) and iii) $\mu_{\pi^-} = -m_\pi$ ($\mu_{\pi^+} = m_\pi$). For the sake of comparison, the BG results are also plotted for case i). We can see that in this case, the electron chemical potential increases until it reaches the pion mass value and then it is kept constant, due to the fact that the negative pions start to replace the electrons. In this case, zero charged and positive pions are almost inexistent. The difference between the BG and non-extensive statistics resides again at very low densities, where the appearance of pions is slightly quenched if Tsallis statistic is used, as can be seen from the top graphs in Fig 8. For this study $q = 1.14$. From Fig 8, one sees that, in these cases, the amount of pions decreases as the density of the system increases, contrary to what happens in case i). Hence, the pion contribution to the system can vary considerably, depending on the way its chemical potential is fixed. These differences in the constitution of the star at low densities are very important because they may influence the existence of the pasta phase [35] and control neutrino mean free paths through the diffusion coefficients [36], with possible consequences in the protoneutron star cooling mechanism.

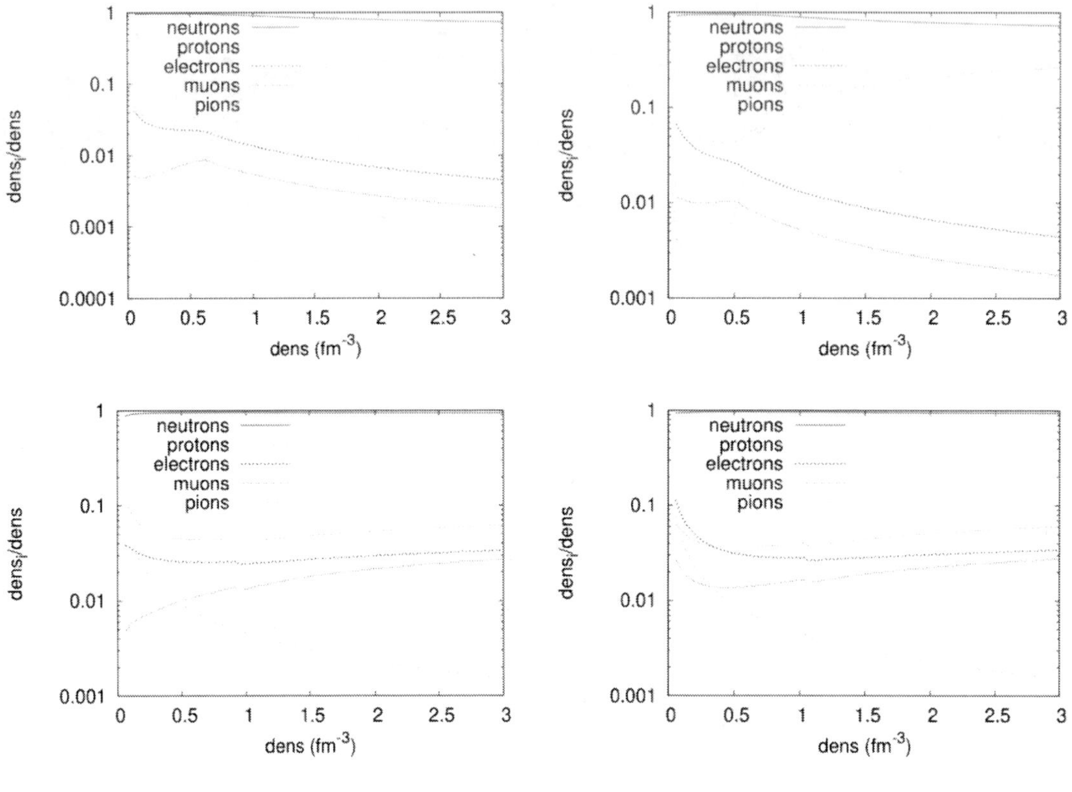

Figure 8. Particle fractions as a function of the baryonic density at $\tau = 20$ MeV. Top graphs show the results for i) $\mu_{\pi^-} = \mu_e$ for left) BG and right) Tsallis statistics ($q = 1.14$). Bottom graphs are shown for Tsallis statistics ($q = 1.14$) for left ii)$\mu_{\pi^-} = m_\pi$ and right iii) $\mu_{\pi^-} = -m_\pi$.

We now turn our attention to the different EOS obtained so far. In Fig. 9 we plot the EOS for both systems studied in the present work, i.e., with and without pions. The results for the BG statistics is shown for comparison. The curves are very similar, but one can check that the inclusion of pions makes the EOS softer in both statistics as far as case i) for the definition of the pion chemical potential is used. In the other situations (case ii) and iii)) the EOS are practically coincident. We conclude that the enforcement of charge neutrality and β-equilibrium stability conditions wash out completely the hardening effect of the pions on the EOS. The difference in the values of the pressure and energy density as compared with the ones presented in Fig. 5 are also caused by the inclusion of the leptons and the stellar matter conditions.

The Tolman-Oppenheimer-Volkof (TOV) equations [37] are the differential equations for the structure of a static, spherically symmetric, relativistic star in hydrostatic equilibrium. When solving the TOV equations we get the results for some of the macroscopic characteristics of the star, particularly its maximum mass and its corresponding radius. With the EOS obtained in the present work, all the results lie around $0.7M_\odot$ and 8 Km for the maximum stellar masses and respective radii. Those are the well known values obtained with EOS described by a free Fermi gas and reflect the fact that the nuclear interaction is not taken into account. These values for the stellar masses are well below

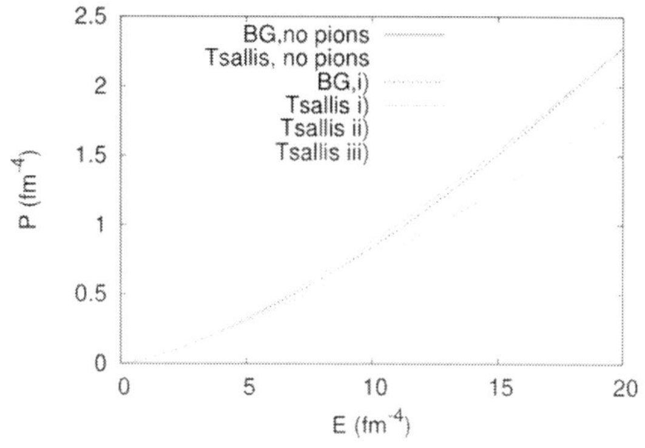

Figure 9. EOS for all cases investigated in the present work.

the expected value of 1.4 or even 2 $M_\odot$ and is in accordance with the analysis made above, where it was evidenced that in this situation both BG and Tsallis statistics yield similar pressures from the hadronic contribution.

V. CONCLUSIONS

In this work we developed the non extensive thermodynamics for an ideal quantum gas for both bosons and fermions from the partition function defined here for the first time. Then we showed that the partition function and the thermodynamics derived from it is equivalent to the thermodynamics derived from the entropy proposed by Conroy, Miller and Plastino [21], and also by Cleymans and Worku [19] when $\mu \leq m$. For $\mu > m$ and for fermions some inconsistencies of previous references were addressed, and our result is fully thermodynamically consistent. In the limit of high energies, the partition function is in accordance with that proposed in Ref. [10], and thus the partition function defined here is self-consistent in the sense proposed by Hagedorn.

Some thermodynamical functions are derived from the partition function for different hadronic systems (proton-neutron-lepton system and proton-neutron-pion- lepton systems) in different conditions. Particularly we analyze how pressure and energy densities vary when the entropic index or the chemical potentials vary. A discussion about a discontinuity observed in the first derivatives of the partition function is done for the first time.

We apply our formalism to study the neutron star stability under several conditions. To this end electron and muon gases in the Boltzmann-Gibbs statistics are included in the system. Our results show that a simple thermodynamical description of the neutron star matter within the non extensive statistics cannot explain the star stability as observed in nature, as expected, since nuclear interaction was not included. Nevertheless, we have seen that the Tsallis statistics modifies substantially the particle constitution at low densities and this fact deserves more investigation towards the analysis of the potentialities of a non extensive thermodynamical approach. In [33], where the non-linear Walecka model was used, the equation of state becomes harder for q values larger than one, with the obvious consequence that the maximum stellar mass is higher than in the usual case. This behavior seems to be different from the one we have obtained, since in the present work, the EOS show very little sensitivity to the increase of q. In [38], a simple model for quark stars was used and, for a fixed temperature, the maximum stellar mass was shown to decrease for q values larger than one, a result which is opposite to the one found in [33]. These somewhat contradictory results deserve better investigation. Moreover, many perspectives for improvements in the simple model proposed here can be done, as the inclusion of other hadrons as the Δ particles, for instance, and the description of leptons also by the non extensive statistics with a q value different from the one used for hadrons.

Acknowledgments

This work has been supported by Plan Nacional de Altas Energías (FPA2011-25948), Junta de Andalucía grant FQM-225, Generalitat de Catalunya grant 2014-SGR-1450, Spanish MINECO's Consolider-Ingenio 2010 Programme CPAN (CSD2007-00042), Centro de Excelencia Severo Ochoa Programme grant SEV-2012-0234, by CNPq (Brazil) and FAPESC (Brazil) under project 2716/2012, TR 2012000344, and FAPESP (Brazil) under grant 2013/24468-1. A.D. acknowledges the support from CNPq under grant 305639/2010-2. E.M. would like to thank the Instituto de

Física of the Universidade de São Paulo for their hospitality and support during the completion of parts of this work. The research of E.M. has been supported by the Juan de la Cierva Program of the Spanish MINECO, and by the European Union under a Marie Curie Intra-European Fellowship (FP7-PEOPLE-2013-IEF). D.P.M. acknowledges the warm hospitality at the Universidad de Alicante, where parts of this work were carried out.

---

[1] C. Tsallis, J. Stat. Phys. 52 (1988) 479.
[2] I. Bediaga, E.M.F. Curado e J.M. de Miranda, Physica A 286 (2000) 156.
[3] R. Hagedorn, Nuovo Cimento Suppl. 3 (1965) 147.
[4] B. Abelev *et al.* (ALICE Collaboration), arXiv:1205.5724v1 (2012).
[5] B. Abelev *et al.* (ALICE Collaboration), Eur. Phys. J. C 72 (2012) 2183.
[6] K. Aamodt *et al.* (ALICE Collaboration), Eur. Phys. J. C 71 (2011) 1655.
[7] V. Khachatryan *et al.* (CMS Collaboration), JHEP 05 (2011) 064.
[8] G. Aad *et al.* (ATLAS Collaboration), Nucl. Phys. B 850 (2011) 387-444.
[9] R. Hagedorn, Lect. Notes Phys. **221**, 53 (1985).
[10] A. Deppman, Physica A 391 (2012) 6380.
[11] N. O. Agasian, Phys. Lett. B **519**, 71 (2001).
[12] W. Broniowski, W. Florkowski and L. Y. .Glozman, Phys. Rev. D **70**, 117503 (2004).
[13] E. Megias, E. Ruiz Arriola and L. L. Salcedo, Phys. Rev. Lett. **109**, 151601 (2012).
[14] E. Megias, E. Ruiz Arriola and L. L. Salcedo, Nucl. Phys. Proc. Suppl. **234**, 313 (2013).
[15] E. Megias, E. Ruiz Arriola and L. L. Salcedo, Phys. Rev. D **89**, 076006 (2014).
[16] L. Marques, E. Andrade-II e A. Deppman, Phys. Rev. D 87 (2013) 114022.
[17] I. Sena e A. Deppman, Eur. Phys. J. A 49 (2013) 17.
[18] I. Sena e A. Deppman, AIP Conf. Proc. 1520, (2013) 172.
[19] J. Cleymans e D. Worku, J. Phys. G: Nucl. Part. Phys. 39 (2012) 025006.
[20] A. Deppman, J. Phys. G: Nucl. Part. Phys. **41**, 055108 (2014).
[21] J.M. Conroy, H.G. Miller and A.R. Plastino, Physics Letters A 374 (2010) 45814584.
[22] E. Megias, D.P. Menezes and A. Deppman, arXiv:1312.7134, accepted for publication in Physica A (2014).
[23] D.P. Menezes, A. Deppman, E. Megias and L.B. Castro, arXiv:1410.2264, submitted for publication in Phys. Rev. C.
[24] E. Megias, D. P. Menezes and A. Deppman, arXiv:1407.8044.
[25] A.M. Teweldeberhan, A.R. Plastino and H. Miller, Physics Letters A 343 (2005) 71-78.
[26] J. Cleymans, K. Redlich, Phys. Rev. C 60 (1999) 054908.
[27] J. Beringer et al., Phys. Rev. D 86 (2012) 010001.
[28] J. M. Lattimer and M. Prakash, Science 304, 536 (2004).
[29] W. Keil and H. Janka, Astron. Astrophys. 296, 145 (1995).
[30] D.P. Menezes and C. Providência, Phys. Rev. C 68 (2003) 035804; Phys. Rev. C 69 (2004) 045801;
[31] A.M.S. Santos and D.P. Menezes, Phys. Rev. C 69 (2004) 045803.
[32] B. D. Serot and J. D. Walecka, Adv. Nucl. Phys. 16, 1 (1986); J. Boguta and A. R. Bodmer, Nucl. Phys. A 292, 413 (1977).
[33] G. Gervino, A. Lavagno and D. Pigato, J. Physics: Conference Series 442 (2013) 012065.
[34] N. K. Glendenning, Compact Stars, Springer-Verlag, New-York, (2000).
[35] H. Shen, H. Toki, K. Oyamatsu, and K. Sumiyoshi, Nucl. Phys. A 637, 435 (1998), S.S. Avancini, C.C. Barros, L. Brito, S. Chiacchiera, D.P. Menezes and C. Providncia, Phys. Rev. C 85, 035806 (2012);
[36] M.D. Alloy and D.P. Menezes, Phys. Rev. C 83, 035803 (2011).
[37] R.C. Tolman, Phys. Rev. **55**, 364 (1939); J.R. Oppenheimer and G.M. Volkoff, Phys. Rev. **55**, 374 (1939).
[38] M. Razeira, B.E.J. Bodmann and C.A.Z. Vasconcellos, Int. J. Mod. Phys. D, 16 (2007) 365.

Soft Interactions at High Energies: Amplitudes and Cross-Sections

Errol Gotsman

Department of Particle Physics, School of Physics and Astronomy, Tel Aviv University, Tel Aviv, 69978, Israel

E-mail: gotsman@post.tau.ac.il

Abstract. I briefly specify elements of the GLM model that successfully describes soft hadronic interactions at energies from ISR to LHC. The model is based on a single Pomeron with a large intercept $\Delta_{I\!P} = 0.23$ and slope $\alpha'_{I\!P} = 0.028$, and so provides a natural matching with perturbative QCD. I analyze the elastic, single diffractive and double diffractive amplitudes, and compare the behaviour of the GLM amplitudes to those of other parameterizations. I then summarize the main features and results of the GLM model and compare with those of the Kaidalov-Poghosyan and other models for soft interactions at high energies.

1. Introduction

The recent measurements of the proton-proton cross sections at the LHC at an energy of W = 7 TeV, allows one to appraise the numerous models that have been proposed to describe soft interactions. The classical Regge pole model *à la* Donnachie and Landshoff [1], which provided a reasonable description of soft hadron-hadron scattering upto the Tevatron energy, fails when extended to LHC energies [2]. In addition, it has the intrinsic problem of violating the Froissart-Martin bound [3].

At present there are a number of models based on Reggeon Field Theory that provide a reasonable description of proton-proton scattering data over the energy range from ISR to LHC. I will describe the essential features of the GLM model [4] as an example of a model of this type, before comparing its results with other competing models on the market.

1.1. Basic features of the GLM model

We utilize the simple two channel Good-Walker (GW) [5] model, to account for elastic scattering and for diffractive dissociation into states with masses that are much smaller than the initial energy, and impose the unitarity constraint by requirying that

$$2\,\mathrm{Im}\,A_{i,k}\,(s,b) = |A_{i,k}\,(s,b)\,|^2 + G^{in}_{i,k}(s,b)$$

where, $A_{i,k}$ denotes the diagonalized interaction amplitude and $G^{in}_{i,k}$, the contribution of all non GW inelastic processes.

A general solution for the amplitude satisfying the above unitarity equation is:

$$A_{i,k}(s,b) = i\left(1 - \exp\left(-\frac{\Omega_{i,k}(s,b)}{2}\right)\right) \qquad (1)$$

the opacities $\Omega_{i,k}$ are arbitrary. In the eikonal approximation $\Omega_{i,k}$ are assumed to be real, and taken to be the contribution of a single Pomeron exchange.

GLM parameterize the opacity :

$$\Omega_{i,k}(s,b) = g_i(b)\, g_k(b)\, P(s)$$

where $P(s) = s^{\Delta}$, and $g_i(b)$ and $g_k(b)$ are the Pomeron-hadron vertices given by:

$$g_l\,(b) = g_l\,S_l(b) = \frac{g_l}{4\pi}\, m_l^3\, b\, K_1\,(m_l b)\,.$$

$S_l(b)$ is the Fourier transform of $\frac{1}{(1+q^2/m_l^2)^2}$, where q is the transverse momentum carried by the Pomeron, $l = i,k$. The form of $P(s)$ used by GLM, corresponds to a Pomeron trajectory slope $\alpha'_{I\!P} = 0$. This is compatible with the exceedingly small fitted value of $\alpha'_{I\!P}$, (0.028 GeV^{-2}) and in accord with N=4 SYM.

For the case of $\Delta_{I\!P} \to 0$, the Pomeron interaction leads to a new source of diffraction production with large mass ($M \propto s$), which cannot be described by the Good-Walker mechanism. Taking $\alpha'_{I\!P} = 0$, allows one to sum all diagrams having Pomeron interactions [6, 7]. This is the advantage of such an approach. The GLM model only takes into account triple Pomeron interaction vertices ($G_{3I\!P}$), this provides a natural matching to the hard Pomeron, since at short distances $G_{3I\!P} \propto \alpha_s^2$, while other vertices are much smaller. A full description of the procedure for summing all diagrams (enhanced + semi-enhanced) is contained in [6, 7, 8]. We would like to emphasize that in the GLM model, the GW sector contributes to both low and high diffracted mass, while the non-GW sector contributes only to high mass diffraction ($\log\left(M^2/s_0\right) \approx 1/\Delta_{I\!P}$).

The GLM model has 14 parameters describing the Pomeron and Reggeon sectors. The values of these parameters are determined by fitting to data for σ_{tot}, σ_{el}, σ_{sd}, σ_{dd} and B_{el} in the ISR-LHC range [8]. We find the best fit value for $\alpha_{I\!P} = 0.21$, however, to be in accord with the LHC data we have tuned $\alpha_{I\!P}$ to 0.23. The fitted values for $\alpha'_{I\!P}$ is 0.028 GeV^{-2}, while the triple Pomeron vertex $G_{3I\!P} = 0.03$ GeV^{-1}.

2. Amplitudes

The Good-Walker formalism [5] provides an explicit form for the various elastic and diffractive amplitudes. Until recently most of the comparison of models has been made on the level of cross-sections (which are areas), and only reveal the energy dependence, and do not display other features. Having the behaviour of the various amplitudes as functions of impact parameter (momentum transfer) would be more revealing. Unfortunately, there is a paucity of material available on amplitudes, and most refer only to the elastic amplitude.

In Fig.1 (left panel) we show elastic amplitudes emanating from the GLM model for various energies. We note the overall gaussian shape of the elastic amplitudes for all energies 0.545 $\leq W \leq 57$ TeV, with the width and height of the gaussian growing with increasing energy. For small values of b the slope of the amplitudes decreases with increasing energy. The elastic amplitude (as $b \to 0$) becomes almost flat for W = 57 TeV, where it is still below the Unitarity limit $A_{el} = 1$.

In Fig. 1 (right panel) we show the energy behaviour of the GLM (G-W contribution) of the single diffractive amplitude as a function of impact parameter for different energies. In Fig. 2 (left panel) we display the behaviour of the double diffractive amplitude. A common feature of both diffractive amplitudes is that with increasing energy the peaks broaden and become more peripheral. In Fig. 2 (right panel) we show the elastic, single diffraction and double diffraction amplitudes as functions of b for W = 7 TeV. Note the completely different shapes of the three amplitudes, the elastic amplitude $A_{el}(b)$ is gaussian in shape, while the single

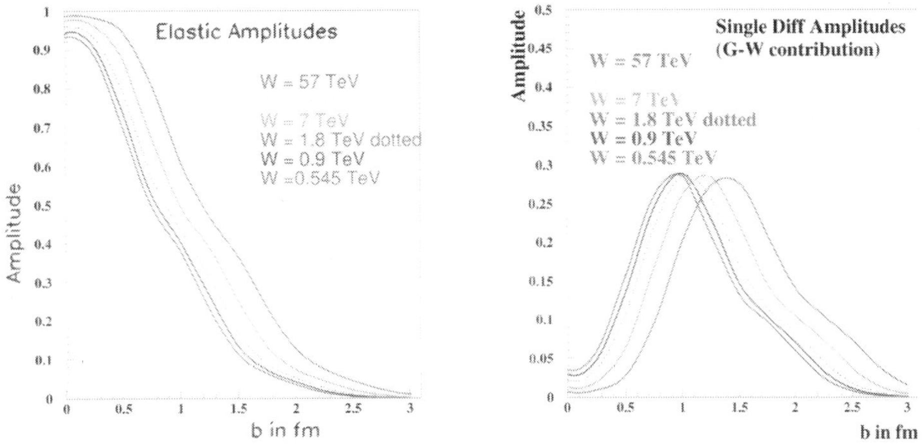

Figure 1. Left panel: GLM elastic amplitudes and Right panel: GLM-GW single diffraction amplitudes.

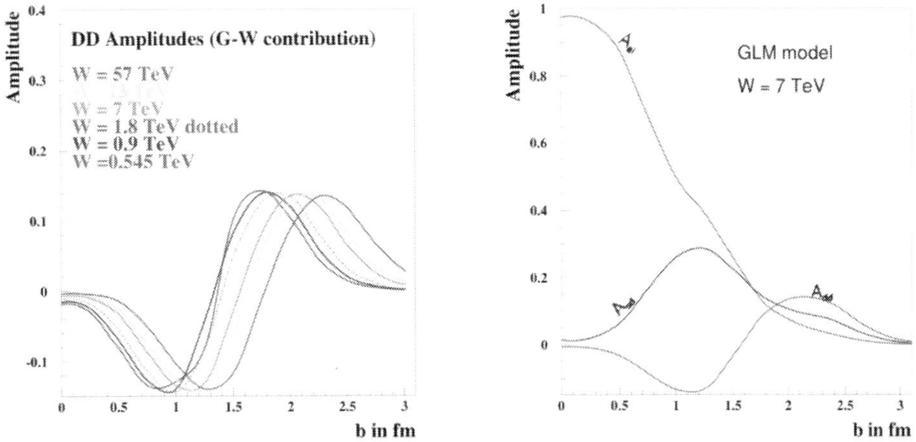

Figure 2. Left panel: GLM double diffractive amplitudes and Right panel: GLM elastic, single diffractive and double diffractive amplitudes for W = 7 TeV.

diffractive amplitude $A_{sd}(b)$ and the double diffractive amplitude $A_{dd}(b)$ are very small at small b . A_{sd} has a peak at 1.25 fm, while A_{dd}'s maximum is at b = 2.15 fm.

The Durham group [9] have attempted to extract the form of the Elastic Opacity directly from the data.They assume that at high energies the real part of the scattering amplitude is very much smaller than the imaginary part, then to a good approximation

$$A(b) = i[1 - \exp(-\Omega(b)/2)]$$

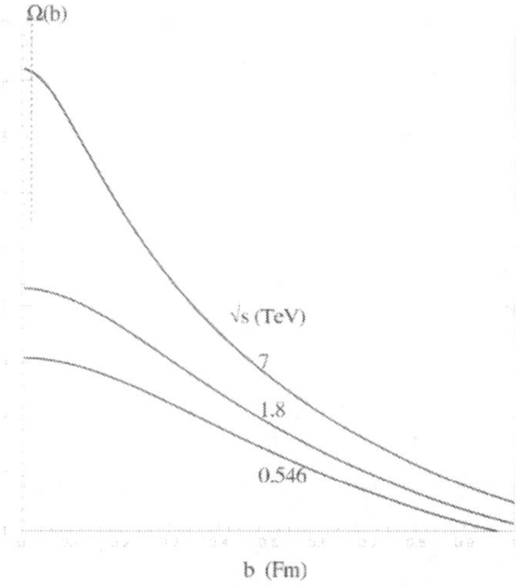

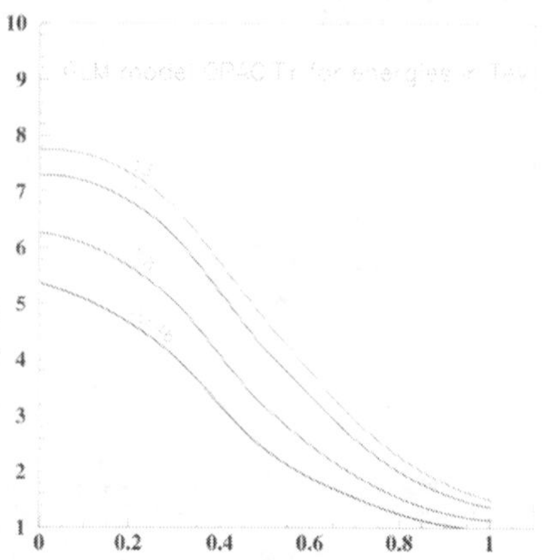

Figure 3. Left panel: The proton opacity $\Omega(b)$ determined directly from the pp $d\sigma_{\rm el}/dt$ data at 546 GeV , 1.8 TeV and 7 TeV data. The uncertainty on the LHC value at $b = 0$ is indicated by a dashed red line. This figure is taken from [9] which should be consulted for details. Right panel: Opacites calculated using the GLM model.

(see Eqn(1)). As $\Omega_{el} = -2ln(1 - A_{el})$, they determine the Opacity directly from the data since

$$\mathrm{Im}A(b) \;=\; \int \sqrt{\frac{d\sigma_{\rm el}}{dt}\frac{16\pi}{1+\rho^2}}\; J_0(q_t b)\; \frac{q_t dq_t}{4\pi},$$

where $q_t = \sqrt{|t|}$ and $\rho \equiv \mathrm{Re}A/\mathrm{Im}A$. Their results are shown in Fig. 3 (left panel) .

The Durham group [9] find that at $Sp\bar{p}S$ and Tevatron energies the Opacity distributions have appoximately a Gaussian form. The analogous GLM model results are shown in Fig. 3 (Right panel), are in agreement with [9] regarding the shape of $\Omega_{el}(b)$, and in addition suggest that this is also true for the LHC energies. GLM find that with increasing energy, the intercept of the Opacity at b =0 increases, while the slope at small b decreases.

Kopeliovich, et al [10] have calculated the proton-proton elastic amplitude within the framework of a two scale dipole model. We show their result in Fig. 4 (Left panel). As well as that of Ferreira, Kodama and Kohara [11] who have recently made a detailed study of the proton-proton elastic amplitude for center of mass energy W = 7 TeV, based on the Stochastic Vacuum Model.

In Fig. 4 (Left panel) we compare the GLM, KPPS and FKK elastic amplitudes at W = 7 TeV as a function of the impact parameter. Although the shapes are similar, the KPPS and FKK amplitudes have lower intercepts at b = 0. If we normalize the FKK amplitude to the GLM value at b = 0, we note that the amplitudes which are gaussian in shape, have very similar behaviour as a function of the impact parameter. In Fig.4 (Right panel) we show the single diffraction amplitude as given by the DIPSY Monte Carlo [12] (dashed line) at W = 1.8 and 14 TeV. This includes contributions both from the Good-Walker sector and enhanced and semi-enhanced sector. The full line is the GLM amplitude which only contains the Good-Walker contribution. Note, although the amplitudes for the same value of W, peak at the same value of b, the DIPSY amplitudes are broader and higher, due to the additional enhanced contributions. For historical purposes we mention that the impact parameter behaviour of our

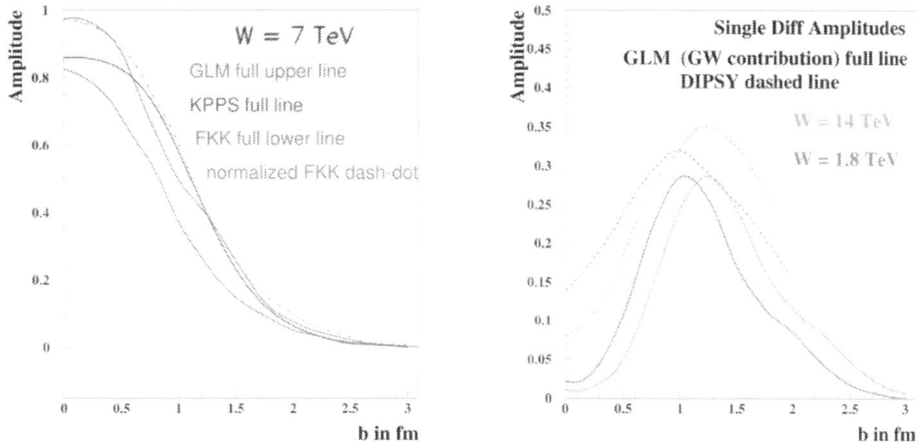

Figure 4. Left panel: Comparison of the elastic amplitude at 7 TeV determined by FKK [11], KPPS [10] and GLM. Right panel: Comparison of single diffraction amplitude, dashed line DIPSY [12] (which includes enhanced and semi-enhanced contributions) and full line GLM (only GW contribution)

diffractive amplitudes are in accord with the estimates of Mietten and Pumplin [13] made over 35 years ago.

3. Experimental Data and GLM results

Our results for σ_{inel}, σ_{sd} and σ_{dd}, are contained in Fig. 5 which is taken from the talk given by Orlando Villalobos Baillie (for the ALICE collaboration) (see reference [14]), where the experimental data, our results and the results of other models are displayed.

The comparison of our results with experimental data for σ_{tot}, σ_{el} and for B_{el} is shown in Fig. 6.

To summarize our results at high energy, we obtain a very good reproduction of TOTEM's values for σ_{tot} and σ_{el}. The quality of our good fit to B_{el} is maintained. As regards σ_{inel}, our results are in accord with the higher values obtained by ALICE [15] and TOTEM [16]; ATLAS [17] and CMS [18] quote lower values with large extrapolation errors, see [19].

There are also recent results at $W = 57$ TeV by the Auger Collaboration [20] for σ_{tot} and σ_{inel}. In Table 1 we compare the experimental results at W = 7 and 57 TeV with the predictions the GLM model.

4. Alternative Models

4.1. Kaidalov-Poghosyan Model

Kaidalov and Poghysan [21] in 2009 suggested a model based on Reggeon calculus, which predicted the soft diffractive results expected at the LHC. Their model took into account all possible non-enhanced absorptive corrections to three Reggeon vertices and loop diagrams. They apply AGK rules for calculating the discontinuity of the matrix element, and a generation of the optical theorem for the case of multi-Pomeron exchange. They utilize a single Pomeron with intercept $\Delta_{I\!P} = 0.12$ and slope $\alpha'_{I\!P} = 0.22\ GeV^2$, as well as secondary Regge poles. The KP model forms the basis of the Monte Carlo program used by the ALICE collaboration to analyze their soft scattering data [19].

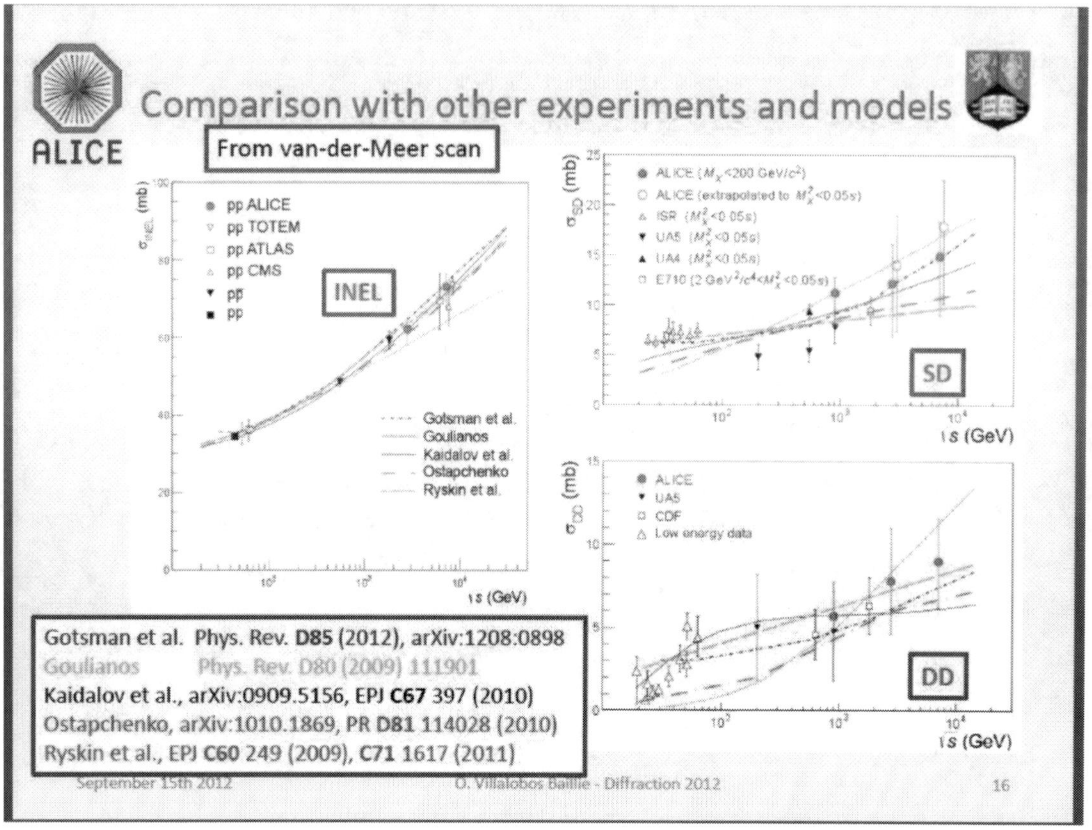

Figure 5. Comparison of Models with LHC data from Villalabos Ballie's talk at Diffraction 2012. [14]

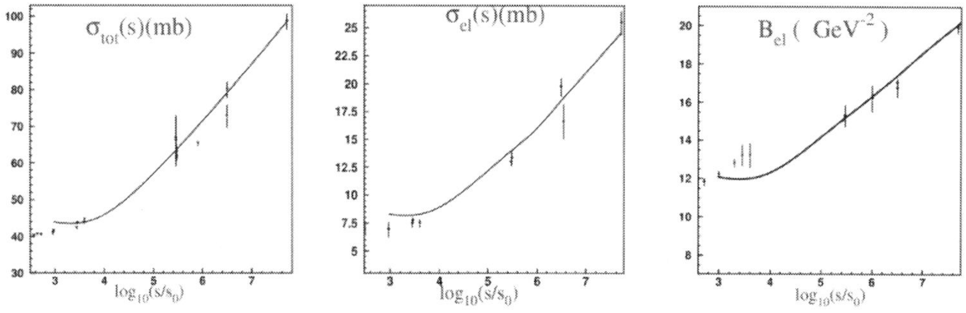

Figure 6. The GLM results compared to data for σ_{tot}, σ_{elas} and B_{elas}

The predictions of Kaidalov-Poghsyan [21] appear in Table 2, in the column KP.

4.2. Other Models on the Market

There are several models on the market today that reproduce the LHC experimental results. The most promising of these are summarized here, and their results are compared with those of GLM [4] in Table 1.

Table 1. Comparison of the values obtained from the GLM model with experimental results at W = 7 and 57 TeV.

W	σ_{tot}^{model}(mb)	σ_{tot}^{exp}(mb)	σ_{el}^{model}(mb)	σ_{el}^{exp}(mb)
7 TeV	98.6	TOTEM: 98.6 ±2.2	24.6	TOTEM: 25.4±1.1

W	$\sigma_{in}^{model}(mb)$	σ_{in}^{exp}(mb)	$B_{el}^{model}(GeV^{-2})$	$B_{el}^{exp}(GeV^{-2})$
7 TeV	74.0	CMS: $68.0 \pm 2^{syst} \pm 2.4^{lumi} \pm 4^{extrap}$ ATLAS: $69.4 \pm 2.4^{exp} \pm 6.9^{extrap}$ ALICE: $73.2\ (+2./-4.6)^{model} \pm 2.6^{lumi}$ TOTEM: $73.5 \pm 0.6^{stat} \pm 1.8^{syst}$	20.2	TOTEM: 19.9±0.3

W	$\sigma_{sd}^{model}(mb)$	σ_{sd}^{exp}(mb)	σ_{dd}^{model}(mb)	σ_{dd}^{exp}(mb)
7 TeV	$10.7^{GW} + 4.18^{nGW}$	ALICE : 14.9(+3.4/-5.9)	$6.21^{GW} + 1.24^{nGW}$	ALICE: 9.0 ± 2.6

W	σ_{tot}^{model}(mb)	σ_{tot}^{exp}(mb)
57 TeV	130	AUGER: $133 \pm 13^{stat} \pm 17^{sys} \pm 16^{Glauber}$
	σ_{inel}^{model}(mb)	σ_{inel}^{exp}(mb)
	95.2	AUGER: $92 \pm 7^{stat} \pm 11^{syst} \pm 7^{Glauber}$

The Durham group's approach for describing soft hadron-hadron scattering [22] is similar to the GLM [4] approach, they include both enhanced and semi-enhanced diagrams. The two groups utilize different techniques for summing the multi-Pomeron diagrams. The Durham Group [23] to be consistent with the TOTEM result [16], have a model, based on a three channel eikonal description, with three diffractive eigenstates of different sizes, but with only one Pomeron. $\Delta_{I\!P} = 0.14$; and $\alpha'_{I\!P} = 0.1$ GeV^{-2}. Which we will refer to as KMR3C.

Recently KMR [24] suggested a two channel eikonal model where the Pomeron couplings to the diffracive eigenstates depend on the collider energy. They have four versions of the model. The parameters of the Pomeron of their "favoured version" Model 4 are: $\Delta_{I\!P} = 0.11$; and $\alpha'_{I\!P} = 0.06$ GeV^{-2}. We refer to this as KMR2C.

Ostapchenko [25] [pre LHC] has made a comprehensive calculation in the framework of Reggeon Field Theory, based on the resummation of both enhanced and semi-enhanced Pomeron diagrams. To fit the total and diffractive cross sections he assumes two Pomerons: (for his solution set C) "Soft Pomeron" $\alpha^{Soft} = 1.14 + 0.14t$ and a "Hard Pomeron" $\alpha^{Hard} = 1.31 + 0.085t$. His results are quoted in Table 2, in the column Ostap(C).

Ciesielski and Goulianos have proposed an "event generator" [26] which is based on the MBR-enhanced PYTHIA8 simulation. In Table 2 their results are denoted by MBR.

5. Conclusions

We [4] have succeded in building a model for soft interactions, which provides a very good description of all high energy data, including the LHC measurements. The model is based on a Pomeron with a large intercept ($\Delta_{I\!P} = 0.23$) and very small slope ($\alpha'_{I\!P}$ =0.028). We find no need to introduce two Pomerons: i.e. a soft and a hard one. The Pomeron in our model provides a natural matching with the hard Pomeron in processes that occur at short distances. The qualitative features of our model are close to what one expects from N=4 SYM [6, 7], which

Table 2. Comparison of results of the different models for $W = 1.8$, 7 and 14 TeV.

W = 1.8 TeV	GLM	KMR3C	KMR2C	Ostap(C)	BMR*	KP
$\sigma_{\text{tot}}(mb)$	79.2	79.3	77.2	73.0	81.03	75.0
$\sigma_{\text{el}}(mb)$	18.5	17.9	17.4	16.8	19.97	16.5
$\sigma_{SD}(mb)$	11.27	5.9(LM)	2.82(LM)	9.2	10.22	10.1
$\sigma_{DD}(mb)$	5.51	0.7(LM)	0.14(LM)	5.2	7.67	5.8
$B_{el}(GeV^{-2})$	17.4	18.0	17.5	17.8		
W = 7 TeV	GLM	KMR3C	KMR2C	Ostap(C)	BMR	KP
$\sigma_{\text{tot}}(mb)$	98.6	97.4	96.4	93.3	98.3	96.4
$\sigma_{\text{el}}(mb)$	24.6	23.8	24.0	23.6	27.2	24.8
$\sigma_{SD}(mb)$	14.88	7.3(LM)	3.05(LM)	10.3	10.91	12.9
$\sigma_{DD}(mb)$	7.45	0.9(LM)	0.14(LM)	6.5	8.82	6.1
$B_{el}(GeV^{-2})$	20.2	20.3	19.8	19.0		19.0
W = 14 TeV	GLM	KMR3C	KMR2C	Ostap(C)	BMR	KP
$\sigma_{\text{tot}}(mb)$	109.0	107.5	108.	105.	109.5	108.
$\sigma_{\text{el}}(mb)$	27.9	27.2	27.9	28.2	32.1	29.5
$\sigma_{SD}(mb)$	17.41	8.1(LM)	3.15(LM)	11.0	11.26	14.3
$\sigma_{DD}(mb)$	8.38	1.1(LM)	0.14(LM)	7.1	9.47	6.4
$B_{el}(GeV^{-2})$	21.6	21.6	21.1	21.4		20.5

is the only theory that is able to treat long distance physics on a solid theoretical basis.

6. Acknowledgments

I would like to thank Boris Kopeliovich and Gosta Gustafson for supplying the numerical values of their amplitudes, and my colleagues and friends Evgeny Levin and Uri Maor, for a fruitful collaboration.

References

[1] Donnachie A and Landshoff P V 1992 *Phys. Lett.* **296** 222
[2] Donnachie A and Landshoff P V 2011 *"Elastic Scattering at the LHC"*, arXiv:hep-ph/1112.2485
[3] Froissart M 1961 *Phys. Rev.* **123** 1053
Martin A 1966 *Nuovo Cimento* **42A** 930
Martin A 2008 *"Can one improve the Froissart bound?"*, arXiv:hep-ph/0812.0680v3
[4] Gotsman E, Levin E and Maor U 2012 *Phys. Lett.* B **716** 425 and references quoted there
[5] Good M L and Walker W D 1960 *Phys. Rev.* **120** 1857
[6] Gotsman E, Levin E and Maor U 2011 *Eur. Phys. J.* C **71**, 1553
[7] Gotsman E, Levin E, Maor U and Miller J S 2008 *Eur. Phys. J.* C **57**, 689
[8] Gotsman E, Levin E and Maor U 2012 *Phys. Rev.* D **85**, 094007
[9] Ryskin M G, Martin A D and Khoze V A 2012 *Eur. Phys. J.* C **72** 1937
[10] Kopeliovich B Z, Potoshnikova I K , Povh B and Schmidt I 2007 *Phys.Rev.* D **76** 044020
[11] Ferreira E, Kodama T and Kohara A K 2013 *Eur. Phys. J.* C **73** 2326
[12] Flensburg C and Gustafson G 2010 *JHEP* **1010** 014
[13] Miettinen H I and Pumplin J 1978 *Phys. Rev.* D **18** 1696
[14] Villalobos Baillie O, talk at Diffraction 2012, *"Measurements of Difffractive and Inelastic Cross Section with ALICE at the LHC "* Lanzarote September 2012
[15] Poghosyan M G 2011 *et al* [ALICE Collaboration] *J. Phys. G* **38** 124044
[16] Ferro F [TOTEM Collaboration] 2011 *Europhys. Lett.* **96** 21002
[17] Aad G *et al.* 2011 [ATLAS Collaboration] *Nature Commun.* **2** 463
[18] Chatrchyan S *et al.* [CMS Collaboration] 2013 *Phys. Lett.* B **722** 5

Journal of Physics: Conference Series **607** (2015) 012008 doi:10.1088/1742-6596/607/1/012008

[19] Revol J P 2012 " *Proton-proton cross sections with multiplicities with ALICE at the LHC*" talk at the Low-x Meeting Phapos Cyprus, June 27- July 1
[20] Abreu P *et al.* [Auger Collaboration] 2012 *Phys. Rev. Lett.* **109** 062002
[21] Kaidalov A B and Poghosyan M G 2009 "*Description of soft diffraction in the framework of reggeon calculus: Predictions for LHC*" arXiv:[hep-ph] 0909.5156
[22] Ryskin M G, Martin A D and Khoze V A 2011 *Eur. Phys. J.* C **71** 1617
[23] Ryskin M G, Martin A D and Khoze V A 2012 *Eur. Phys. J.* C **72** 1937
[24] Khoze V, Martin A D and Ryskin M G 2013 *Eur. Phys. J.* C **73** 2503
[25] Ostapchenko S 2010 *Phys. Rev.* D **81** 114028
[26] Ciesielski R and Goulianos K 2012 "*MBR Monte Carlo Simulation in PYTHIA8*" arXiv:[hep-ph] 1205.1446

Dedicated to A.B. Kaidalov's memory

A-DEPENDENCE OF DIFFRACTION DISSOCIATION PHENOMENON IN NEUTRINO INTERACTIONS WITH NUCLEI

O.K. Egorov

Institute for Theoretical and Experimental Physics, Moscow, Russia

E-mail: egorov@itep.ru

Abstract. We present results on A-dependence of diffraction dissociation (DD) phenomenon in interactions of muonic neutrino with nuclei at various values of momentum transfer. Calculations were done within the two-gluon exchange model by Nikolaev-Zakharov-Zoller (NZZ). Comparison with experimental data obtained in interactions of muonic neutrino with photoemulsion nuclei is carried out. All data are presented as ratio of the DD cross section to the deep inelastic cross section.

Introduction

When working on analysis of electroproduction data (the experiment E-665/FNAL/) on muon interactions with xenon nuclei the theoretical group leading by N.N. Nikolaev noticed [5] that the contribution of diffractive dissociation (DD) processes on heavy nuclei can reach at definite values of kinematical variables one half of the total deep inelastic cross section. However the authors of the experiment E-665 give different values for this ratio. Namely: $R = \sigma_{DD}/\sigma_{DIS} = 0.12 \pm 0.02$ for μD and 0.18 ± 0.03 for μXe interactions [5, 8]. It is essentially less compared to the estimate of this ratio according to the NZZ model.

The experimental investigation of the diffraction dissociation phenomenon in the photo and electro production became possible only after start-up of the electron-proton accelerator HERA. In particular, investigation of this phenomenon by collaborations H1 and ZEUS is presented in papers [6, 7]. The ratio R given in these papers is $\sigma_{DD}/\sigma_{DIS} \approx 0.1$ being consistent with the NZZ estimate within experimental errors.

Diffractive dissociation calculation procedure

According to [1] the deep inelastic scattering (DIS) of muon neutrino on proton is described by diagram of Fig.1. The similar diagram describes the neutrino-nucleus scattering.

The diffraction dissociation process can be described with this diagram in the DD kinematics (Fig.2) when the events are characterized by large (pseudo)rapidity gap (LRG) $\Delta\eta$ between the

Journal of Physics: Conference Series **607** (2015) 012009 doi:10.1088/1742-6596/607/1/012009

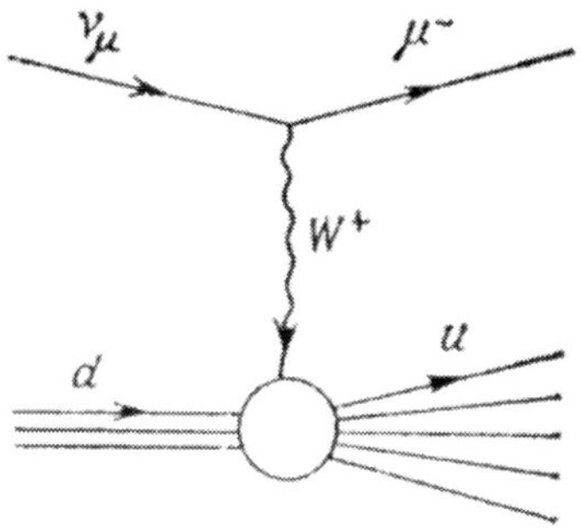

Figure 1. Deep inelastic scattering diagram

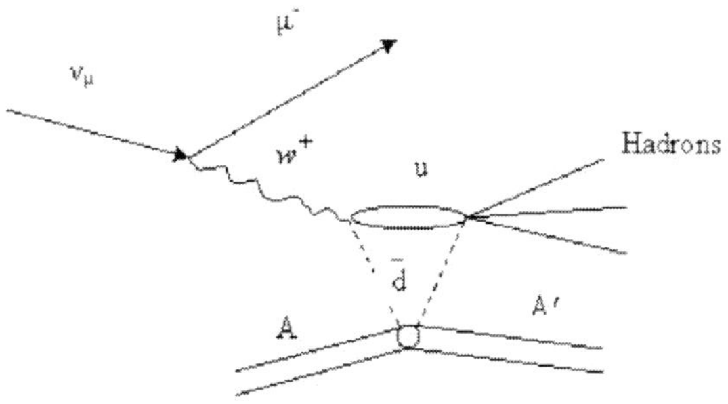

Figure 2. Deep inelastic scattering diagram in the DD kinematics (dashed lines mean the diffractive exchange)

recoil proton (nucleus) and hadrons produced due to the neutrino diffraction dissociation:

$$\Delta\eta = \log(W^2/Q^2) = \log(1/x) - \log(M^2/Q^2) \ , \tag{1}$$

where W^2 is the squared total DIS energy, Q^2 is the squared momentum transfer to the hadron system, and M^2 is the diffractively produced hadron mass squared.

The DD calculation procedure is presented in detail in refs.[2]-[5]. The gap $\Delta\eta$ is about to for $W^2/Q^2 \sim 100$ and $M^2/Q^2 \sim 0.1$, corresponding to the Bjorken variable x about ~ 0.1. The DD process at $x \leq 0.1$ is rather good described by two-gluon exchange (see [2]-[5]) as can be seen from Fig3. The quark exchange contributions (multiplied by 10 at the figure) can be neglected in first approximation.

Results and discussion

Our calculations of the A-dependence for neutrino-nucleus interactions using the NZZ model are presented at Fig.4. The ratio $R = \sigma_{DD}/\sigma_{DIS}$ is estimated for protons, nitrogen nuclei (gelatin, a part of photoemulsion), bromine and silver nuclei. Calculation accuracy is about $\sim 5\%$. Gluon mass is taken as 150 Mev/c^2. The standard spherical form of nucleus with diffused edge is assumed ($R = 1.1$ fm $\cdot A^{1/3}$ and the Fermi momentum 250 MeV/c). We used the following experimental data for νA interactions. The experiment E-564 /FNAL/ gives $\sigma_{DD}/\sigma_{DIS} = 0.29 \pm 0.09$ [9] for AgBr crystals. Though the average Q^2 value is several times higher than 1, one can believe that calculation is consistent with the E-564 data. However more than half of this ratio is due to the incoherent DD.

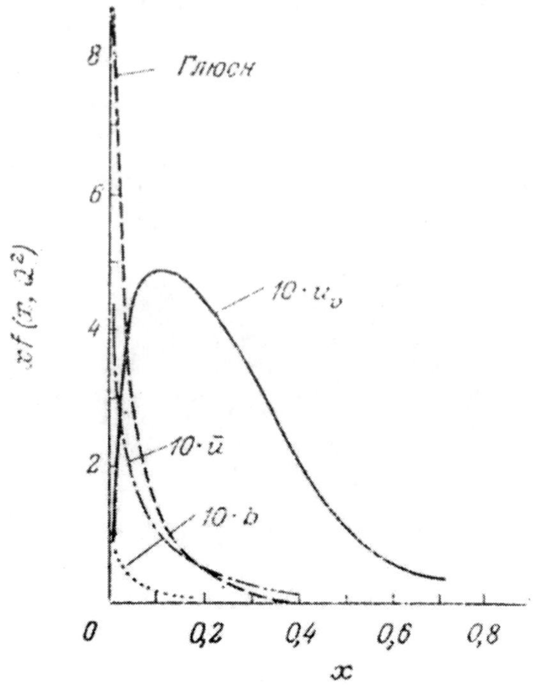

Figure 3. Structure function dependence on the Bjorken variable x

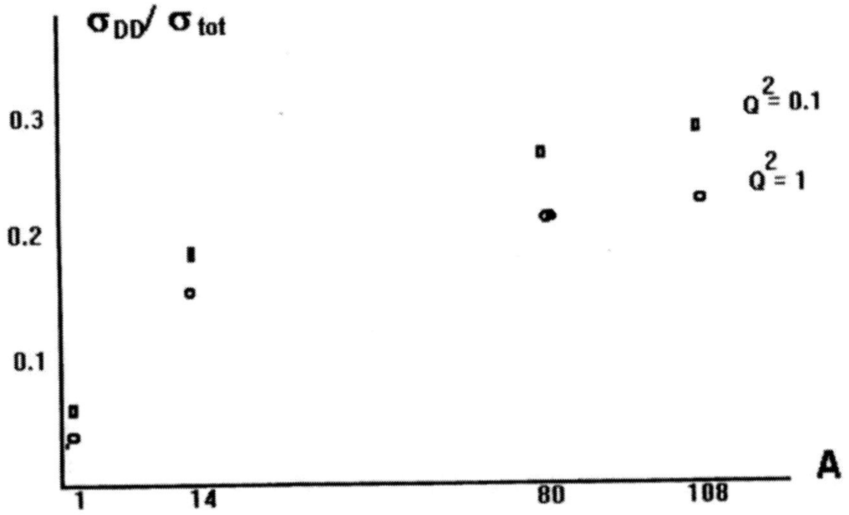

Figure 4. A-dependence for νA interactions

Statistics in the experiment E-128 /Protvino/ is two times larger than in E-564 (670 events) but it gives only the upper bound for the ratio $\sigma_{DD}/\sigma_{DIS} \leq 0.53 \pm 0.07$ [10]. Notice that in this experiment instead of the Bjorken scaling variable $x = Q^2/2M\nu$ a new variable x' is used which, in contrast to x, takes into account nonzero proton mass:

$$x' = x/(1 + m_N^2 x^2/Q^2) \ . \tag{2}$$

Here m_N is a nucleon mass and ν = is energy transferred to hadrons. Comparison of x and x' distributions was carried out at the SCAT bubble chamber [10]. An analysis of those distributions shows that substitution$x$ to x' results in about 20% increase of event number in the $0 - 0.1$ region, i.e. a number of candidates to the DD effect at freon bromide increased at 20%. This estimate id dane without taking into account for lack of events at very low x'. So our estimate of relative contribution of the DD cross section to the deep inelastic cross section at freon bromide increases from 0.17 to 0.21 when substituting x to x'. Account for low effectiveness of detection in small x region gives the ratio 0.31. Just the similar correction resulted in so high upper bound in the experiment E-128.

Conclusions

In summary it is worth noting that physics at small x is not so serene as it seems to the authors of [1]. In particular, it is noted in ref.[9] that the large part of the DD events are related to the noncoherent DD, i.e. those are not "white stars". The author hopes that the OPERA experiment which has already a statistics of many thousand neutrino interactions will be able to elucidate the situation [11].

The author is grateful to his colleagues involved to the E-564 and E-128 experiments and looks forward to efforts of his OPERA colleagues in obtaining new DD estimates.

References

[1] G. Kane. *Modern Elementary Particle Physics*, Addison-Wesley Pub. Com. Inc. 1987
[2] N.N. Nikolaev and B.G. Zakharov. *Z. Phys. C* 1991 **49** 607.
[3] N.N. Nikolaev and B.G. Zakharov. *Z. Phys. C* 1992. **53** 331.
[4] N.N. Nikolaev, B.G. Zakharov and V.R. Zoller. *Z. Phys. A* 1995 **351** 435.
[5] N.N. Nikolaev, B.G. Zakharov and V.R. Zoller. Preprint JÜLICH, KFA-IKP(TH)-1994-13
[6] C. Adloff, S. Aid, M. Anderson et al. Preprint # 97-009 (DESY, Hamburg, 1997).
[7] J. Breitweg, M. Derrick, D. Krakauer et al. *Phys. Lett. B* 1998 **421** 368.
[8] M.R. Adams, S. Aid, P.L. Anthoni et al. FERMILAB-Pub-95/396-E
[9] O.K. Egorov *Yadernaya Fizika* 2002 **65** 881.
[10] O.K. Egorov, V.A. Ryabov. *Yadernaya Fizika* 2004 **67** 2190.
[11] R. Acquafredda et al. 2009 JINST **4** P04018

Phenomenology of high multiplicity events

E S Kokoulina on behalf of SVD Collaboration

JINR, VBLHE, Dubna, Moscow region, Russian Federation, 141980 and Gomel state technical university, Belarus, 246000

E-mail: kokoulin@sunse.jinr.ru

Abstract. Study of high multiplicity events in proton-proton (nucleus) interactions is carried out at U-70 accelerator of IHEP, Protvino. These events are extremely rare. Usually, MC codes underestimate topological cross sections in this region. Gluon dominance model has been offered to describe them. Some collective phenomena can be observed in high multiplicity events. SVD-2 Collaboration has obtained the evidence of pion condensate formation at high total multiplicity. Our future program is aimed at studying soft photon yield puzzle at U-70 and Nuclotron facility at JINR, Dubna.

1. Introduction

Multiparticle production remains an actual theme of modern high energy physics. The SVD Collaboration studies the proton-proton and proton-nucleus interactions at U-70 accelerator of IHEP, Protvino [1] (experiment E-190) with high number of secondary particles

$$p + p(A) \to 2N + N_\pi, \tag{1}$$

where N – nucleon, N_π – number of pions. The energy of the proton beam is equal to 50 GeV, the average charged multiplicity $\overline{N}_{ch}$ is equal to 5.45. Pions are hadrons copiously produced at this energy. We study the region of high charged (N_{ch}) and total (N_{tot}) multiplicity considerably more than average one. The kinematical limit is defined from a condition of the conversion of all kinetic energy of colliding protons into pions

$$N_{thresh} \simeq (\sqrt{s} - 2m_p)/m_\pi. \tag{2}$$

At 50 GeV/c the kinematical limit is equal approximately to 59 pions. The experiment is carried out at SVD (Spectrometer with Vertex Detector) setup located at U-70 accelerator of IHEP, Protvino [1, 2]. Its main elements are a hydrogen target, a silicon vertex detector, a drift tube tracker, a magnetic spectrometer (proportional chambers and a large magnet) and an electromagnetic calorimeter [3]. The setup registers charged particles and photons. The high charged multiplicity events are extremely rare. The scintillator hodoscope (a high multiplicity trigger) has been manufactured to suppress the small multiplicity events and to register events with multiplicity higher than a given level [4].

It is known that Monte-Carlo event generators are often mistaken in their predictions of topological cross sections in the high multiplicity region [5]. They usually underestimate theirs. The existent models give diverse predictions of high multiplicity behavior [6]. So the experimental and following theoretical studies are necessary. SVD Collaboration has advanced in the measuring topological cross sections of pp interactions [1, 2] with high charged and total multiplicity. Previous Mirabelle data [7] have been renovated for N_{ch} from 10 up to 16 and we added 4 new points from 18 up to 24. Topological cross section at the last observing point, $N_{ch} = 24$, is three order of magnitude lower than it was obtained by the Mirabelle Collaboration at $N_{ch} = 16$.

Neutral particles, photons, are registered by an electromagnetic calorimeter (ECal). Owing to its restricted acceptance it is impossible to restore all neutral pions directly. That is why the original algorithm has been developed to define number of events with certain multiplicity of π^0-mesons [8, 9, 10, 11, 12]. The number of events with given total multiplicity, $N_{tot} = N_{ch} + N_0$, where N_0 is multiplicity of neutral pions, is restored by the same method.

Some collective phenomena are predicted in this region. We are aimed to search for pion condensate or Bose-Einstein condensation (BEC) [13], the peak structure at the angular distributions stipulated to a Cherenkov radiation or a shock wave formation [14] and the study of the anomalous soft photon ($p_T < 100$ MeV) yield [15]. We have gotten preliminary indications at the two-humped structure in angular distribution in events with multiplicity more than average one [16, 17]. We have developed the gluon dominance model (GDM) to describe topological cross sections in high multiplicity region [16, 17]. We have estimated the size of region for the soft photon radiation by GDM [18]. This region exceeds the size of particles slightly.

In section 2 the results of search for BEC is presented. The description of high multiplicity region in the framework of GDM is added in sections 3 and 4. Section 3 demonstrates how GDM improves description of the high multiplicity tail for topological cross sections with taking into account gluon fission. Estimation of the charged exchange contribution in GDM is carried out in section 4. Section 5 states the conclusions.

2. Collective phenomena at high multiplicity

With increasing the total multiplicity of pions their average energy is decreased. Pions are bosons and can form pion condensate at low temperature. M Gorenstein and V Begun had analyzed the conditions of BEC formation in the framework of the ideal pion gas model [13, 19]. They predict the increase of fluctuations of neutral pion number at the growth of total pion multiplicity and the approach to BEC. According to their predictions, the scaled variance, ω determined as the ratio of variance, D, to the mean multiplicity of neutral pions, $\overline{N}_0$, at given total multiplicity, can give the evidence of the BEC formation. In this case an abrupt and anomalous increase of ω is expected closely to the vicinity of the pion condensate point [19]. Monte-Carlo event generators do not reveal the growth of ω. SVD Collaboration has seen this growth, and the evidence of

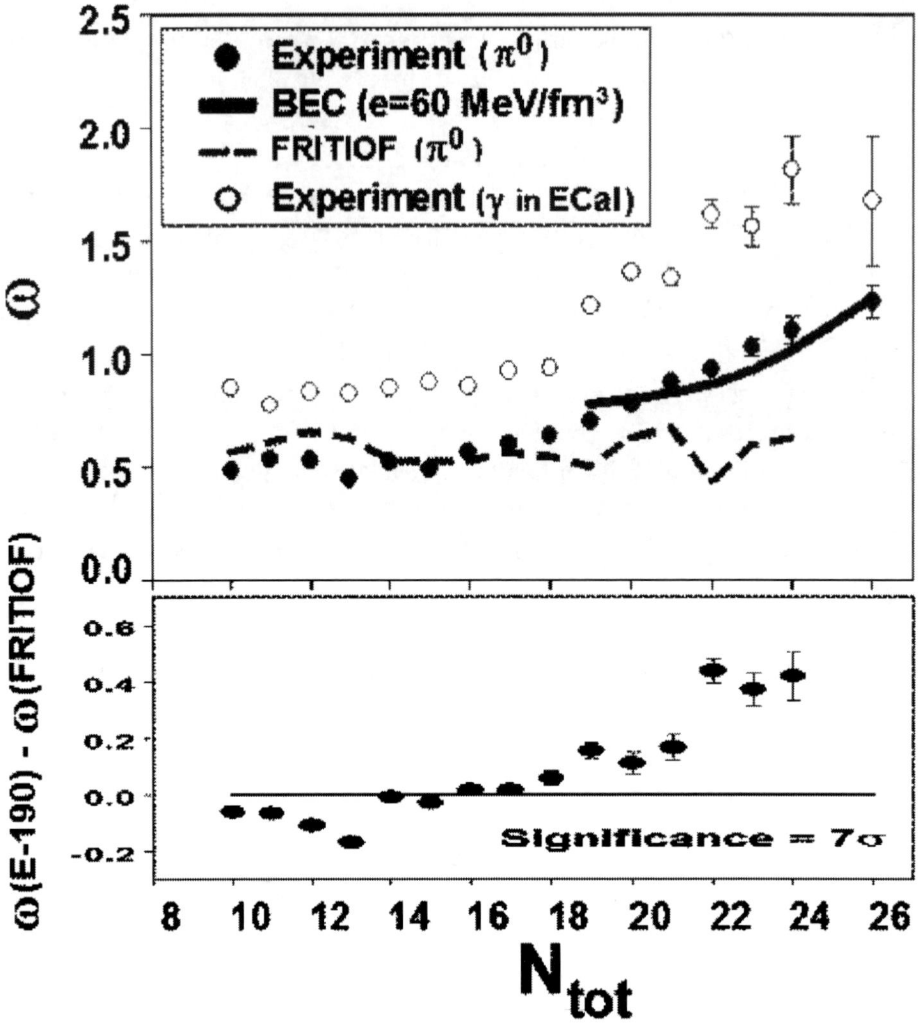

Figure 1. (Top) The measured scaled variance ω versus N_{tot} for π^0-mesons ($\bullet$), photons ($\circ$), MC code FRITIOF7.02 (the dashed curve) and theoretical prediction (solid curve) [13] for the energy density ε =60 MeV/fm^3. $N_{tot} = N_{ch} + N_0$ for π^0-mesons and $N_{tot} = N_{ch} + N_\gamma$ for photons. (Bottom) The difference of experimental and Monte Carlo simulated ω for π^0-mesons [9].

Begun-Gorenstein predictions has obtained [8].

In the top panel of figure 1 the experimental and model values of scaled variances are shown. The experimental values agree well with ω defined by Monte-Carlo simulated events at $N_{tot} < 18$. At the same time we reveal its significant growth, reachable by more than 7 standard deviations at $N_{tot} \simeq 30$ (the bottom panel in figure 1) as opposed to the tendency for the simulated events. This growth has been observed both registered photons ($N_{tot} = N_{ch} + N_\gamma$), and restored neutral pions ($N_{tot} = N_{ch} + N_{ch}$). The theoretical predictions done by V. Begun and M. Gorenstein [13] at a choice of the

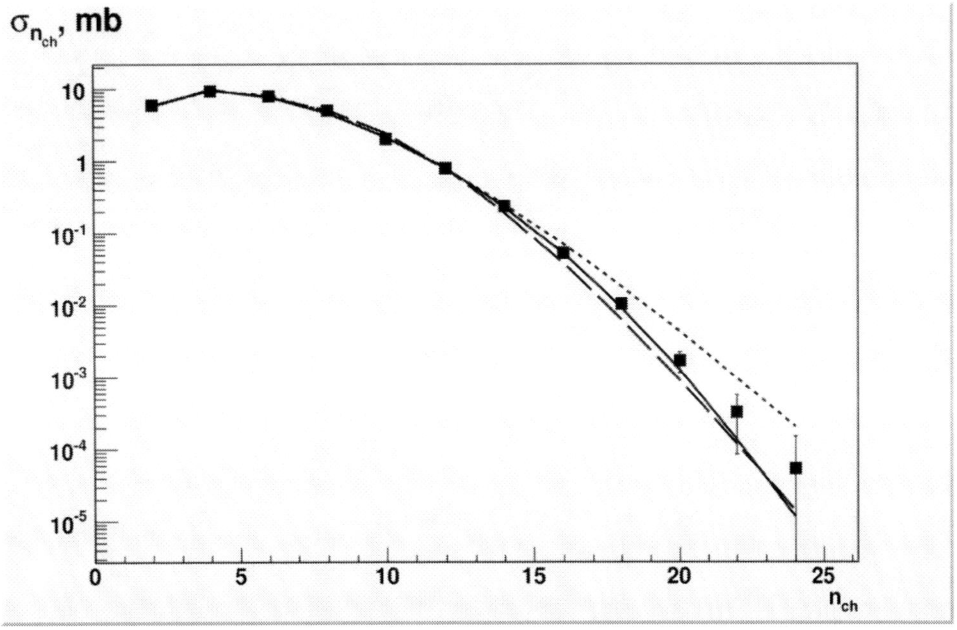

Figure 2. Comparisons data with GDM [21, 22] (solid curve), IHEP model [23] (dotted curve) and NBD distribution [24] (dashed curve).

reasonable values for the size of system, the energy density, the pion number density and others at the high total pion multiplicity area agree to the Bose-Einstein condensate formation since we observe the growth of the scaled variance [9].

The average energy of pion E_π is estimated as

$$E_\pi = (E_{cms} - 2m_N - n_{tot}m_\pi)/n_{tot}, \tag{3}$$

where E_{cms} – energy in cms, m_N – mass of a nucleon, n_{tot} and m_π – number and mass of pions. At $n_{tot} = 30$ the average energy of a pion is equal to $E_\pi = 120$ MeV. The critical point of the pion condensation is determined by statistical physics [20]: $E_{crit} = 3.3(h^{2/3}/m_\pi)\rho^{2/3}$ where ρ is the density of number pions. This density is equal to 0.2 fm^{-3} if the interaction region of two protons is $\simeq 3$ fm and $n_{tot} = 36$ (max observable total multiplicity). In this case the critical energy is equal to $E_{crit} \simeq 100$ MeV. This density is comparable with E_π. In accordance with these estimations BEC can observe at high total pion multiplicity region.

The extended region of multiplicity has been described by few models. In figure 2 the comparison of these models with the measured topological cross sections is shown. The GDM [16, 17, 21, 22] is shown by a solid line, the IHEP model [23] – a dashed line and the negative binomial distribution (NBD) [24] – a dotted line. GDM and IHEP models show the good agreement with data. NBD overestimates our data at high multiplicities.

It is necessary to note that we observe events with number of $n_{ch} = 12$ and $n_0 = 24$ and vice-versa. In this region close to BEC we reveal big fluctuations of neutral pion number. We study multi particle production at higher energy selecting rare high multiplicity events. The investigation of this region is unique, fruitful and promising one.

S Barshay [25] sees an interest connecting between BEC and an anomalous soft photon yield in hadron interactions. He considers that the BEC formation establishes a coherent state when almost all pions have the lowest energy level. In this state pions throw down the excess of energy by the soft photon radiation. To test this hypothesis SVD Collaboration manufactured the electromagnetic calorimeter of soft photons. It consists of 49 BGO crystals and photomultipliers. Now we are preparing for a test run at JINR' Nuclotron [26].

3. GDM

GDM is a modification of the two stage model (TSM) [27, 28]. TSM describes well multiplicity distributions in e^+e^- annihilation in a wide energy region. It is based on the convolution of the QCD quark-gluon cascade of initial quark pair and hadronization stage. In accordance with experimental data the binomial distribution has been chosen for hadronization. The binomial distribution uses the following parameters: the average multiplicity $\overline{n}^h_{q,g}$ and the maximum number of hadrons $N_{q,g}$ resulting from quark or gluon at their passing through this stage. The ratio of quark and gluon parameters, $\alpha = \overline{n}^h_g/\overline{n}^h_q$, reduces the number of parameters at the hadronization stage to three ones. The parameter of second stage $\overline{n}^h_g$ is equal to 1 in all studied energy region. The main result of TSM is the confirmation of fragmentation mechanism of hadronization: one parton fragments for one hadron at hadronization in vacuum.

In pp interactions attend both valency quarks and gluons. At the beginning in GDM valency quarks and some gluons produced secondary hadrons at hadronization. The comparison with experimental data has shown that multi particle production is implemented by gluons called active. Valency quarks are staying in the leading particles as passive spectators. Only in this case the description of topological cross sections becomes possible in this scheme. The hadronization parameter of gluons $\overline{n}^h_g$ grows from 1.5 at 50 GeV/c, U-70, up to 3.3 at 62.2 GeV, ISR. The growth of hadronization parameter is the evidence of the recombination mechanism of hadronization. The second stage occurs in quark-gluon medium not in vacuum.

We used a simple scheme with two stages to describe topological cross sections in pp interactions. At the first stage some active gluons appear. Then they fragment into hadrons. The convolution of two stages describes data well. In the high multiplicity region a little deviation is observed (figure 2). It is known that gluons give branching at high energies. This fission becomes predominating and gives an additional contribution to multiplicity. In GDM it is taken into account in the following way

$$\sigma_n = \alpha_1 \sum_{m_1} \frac{e^{-\overline{m}_1}\overline{m}_1^{m_1}}{m_1!} \binom{m_1 N}{n-2} \left(\frac{\overline{n}^h}{N}\right)^{n-2} \left(1-\frac{\overline{n}^h}{N}\right)^{m_1 N - n + 2} +$$

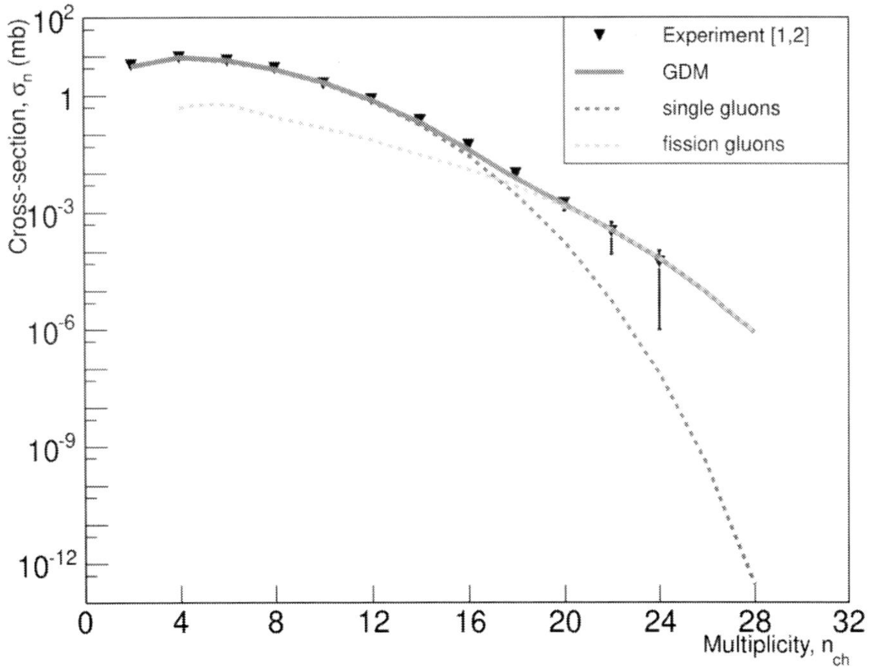

Figure 3. Topological cross section versus charged multiplicity in GDM. The dashed blue line describes the contribution of single sources, the green line – sources consisted of two gluons of fission, the solid red line is the sum both of contributions.

$$+\alpha_2 \sum_{m_2} \frac{e^{-\overline{m}_2}\overline{m}_2^{m_2}}{m_2!} \binom{2m_2N}{n-2} \left(\frac{\overline{n}^h}{N}\right)^{n-2} \left(1-\frac{\overline{n}^h}{N}\right)^{2m_2N-n+2}, \quad (4)$$

where $m_{1,2}$ – multiplicities of single and double gluon sources, $\overline{n}^h$ and N – parameters of hadronization. In this scheme at the second stage the branching gluons give two times more hadrons because they are sources consisted of two gluons (second summand in σ_n). The ratio of two parameters α_1/α_2 is approximately equal to 1.8. The rest parameters do not almost change.

In the double-logarithmic approximation [29] it has been revealed that the emission of two gluon jets formed by the fission process of paternal quark (antiquark) is predominant with the increasing energy and can explain the angle broadening of secondary particles. This fission can give the broadening of topological cross sections at high multiplicity [21, 22, 29]. In figure 3 the comparison of topological cross sections [7, 1] measured with GDM (solid red line) is shown. In accordance with GDM and taking into account the gluon fission the topological cross section consists of two summands. The dash green line presents the contribution appeared from single gluon sources; the second summand (dashed blue line) in (4) describes the contribution responsible for fission gluons. Obviously accounting of the gluon fission improves the description of data a lot.

4. The charged exchange in pp interactions

The experiments show that the charged exchange can be achieved at the scattering of protons off hydrogen or nucleus targets [30]. In this case one of protons gives its charged to a neutral pion and transforms into a neuron

$$p + p \rightarrow p + \pi^+ + n + N_\pi. \quad (5)$$

In (5) N_π – the number of secondary pions, n – a neutron. The estimation of the charge exchange can be carried out in the inelastic channel $2 \rightarrow 2$. In accordance with the Mirabelle data [7] the elastic and inelastic cross sections are equal to $\sigma_{2,el} = 6.90$ mb and $\sigma_{2,inel} = 5.71$ mb accordingly. Their ratio is equal to $r = \sigma_{2,el}/\sigma_{2,inel} = 0.82$. In MGD cross sections $\sigma_{2,el} \sim \exp(-\overline{m})$ because the active gluons are absent. Since we do not know a contribution of the charge exchange, we can use for $\sigma_{2,inel}$ the expression $r \cdot \exp(-\overline{m})$. At the same time $\sigma_{2,inel}$ can be presented as a sum of two summands

$$\sigma_{2,inel} = \sigma^{+ch}_{2,inel} + \sigma^{-ch}_{2,inel}, \quad (6)$$

with a charge exchange (+ch) and without a charge exchange (-ch). GDM estimates the second summand in (6) the following way

$$\sigma^{-ch}_{2,inel} \sim \sum_m \frac{e^{-\overline{m}}\overline{m}^m}{m!} \left(1 - \frac{\overline{n}^h}{N}\right)^{m \cdot N - n + 2}, \quad (7)$$

when active gluons appear without formation of charged particles. Let $P = \sigma_{2,inel}/\sigma^{-ch}_{2,inel}$, then instead $\sigma_{2,inel}$ we use the expression (7) with factor P and from fitting find $P = 2.18$. Hence the coefficient of charge exchange

$$q = \sigma^{+ch}_{2,inel}/\sigma_{2,inel} \cdot 100\%$$

consists of $\simeq 50$ % [31]. This estimation agreed with the experimental data [30]. This value is considerable. It is necessary to have it in mind.

5. Conclusion

In this review the main results of high multiplicity studies have been presented. The region of high multiplicity is unique, promising and fruitful one. We expect new collective phenomena will be found out and will give significant advance for understanding the multi particle production especially the hadronization stage.

6. Acknowledge

I appreciate all participants of SVD-2 Collaboration for their active and fruitful work. I also thank Piskunova Olga for tremendous job on the organization of the workshop devoted to the memory of Professor Alexei Kaidalov.

References

[1] Ryadovikov VN 2012 Topological cross sections in proton-proton interactions at 50 GeV *Phys.Atom.Nucl.* **75**, Issue 3, 315-320.

[2] Kokoulina E S, Nikitin V A, Petukhov Yu P and Kutov A Ya 2012 Proton interactions with high multiplicity *Phys.Atom.Nucl.* **75**, Issue 6, 664-667.

[3] Avdeichikov V V *et al* 2013 Spectrometer with a vertex detector for experiments at the IHEP accelerator *Instrum.Exp.Tech.* **56** 9-31.

[4] Avdeichikov V V *et al* 2011 A trigger of events with a high multiplicity of charged particles at the SVD-2 setup *Instrum.Exp.Tech* **54** 159-168.

[5] Kokoulina E S, Kutov A Ya, Nikitin V A and Popov V V 2011 Analysis of high multiplicity events *Phys.Part.Nucl.Lett.* **8** 855-859. http://dx.doi.org/

[6] Kokoulina E S, Nikitin V A, Petukhov Yu P, Karpov A V and Kutov A Ya 2010 Search for collective phenomena in hadron interactions *Phys.Atom.Nucl.* **73** 2116-2124.

[7] Ammosov V V *et al* 1972 Phys.Lett. B42 (1972) 519-521. http://

[8] Kokoulina E S 2011 Neutral Pion Fluctuations in pp Collisions at 50 GeV by SVD-2 *Prog.Theor.Phys.Suppl.* 193 306-309.

[9] Afonin A G *et al* 2012 Neutral pion number fluctuations at high multiplicity in pp-interactions at 50 GeV *EPJ Web Conf.* **37** 06002. http://dx.doi.org/10.1051/epjconf/20123706002

[10] Ryadovikov V N 2012 Fluctuations of the number of neutral pions at high multiplicity in pp interactions at 50 GeV *Phys.Atom.Nucl.* **75**, 989-998.

[11] Kokoulina E for SVD-2 Collaboration 2013 Evidence for a pion condensate formation in pp interactions at U-70 *PoS ICHEP2012* 259

[12] Kokoulina E S 2012 The evidence for the pion condensate formation in pp interactions at U-70 *PoS Baldin-ISHEPP-XXI* 007

[13] Begun V V and Gorenstein M I 2008 Power Law in Micro-Canonical Ensemble with Scaling Volume Fluctuations *Phys.Rev.* **C**78 024904.

[14] Ulery J G. STAR Collaboration. Are there Mach cones in heavy ion collisions? Three-particle correlations from STAR. 2005 *Int.J.Mod.Phys.* **E**16 2005–10

[15] Chliapnikov P *et al.* 1984 Observation of direct soft photon production in K+ p interactions at 70-GeV/c. *Phys.Let.* 141**B** 276-280.

[16] Kokoulina E S, Kutov A Ya and Nikitin V A 2007 Gluon dominance model and cluster production *Braz.J.Phys.* **37** 785-787.

[17] Kokoulina E S and Kutov A Ya 2008 High-multiplicity study *Phys.Part.Nucl.Lett.* **71**, Nó 9, 1573-1581.

[18] Volkov M K, Kokoulina E S and Kuraev E A 2004 Excitation of Physical Vacuum, *Phys.Part.Nucl.Lett.* Part 5, 16-23.

[19] Begun V V and Gorenstein M I 2007 Bose-Einstein condensation of pions in high multiplicity events. *Phys. Lett.* **B** 653, N2-4, 190-195.

[20] Landau L D and Lifshitz E M 1969 Statistical Physics (Volume **5** of A Course of Theoretical Physics) Pergamon Press.

[21] Kokoulina E S and Nikitin V A 2005 Study of multiparticle production by gluon dominance model *17th International Baldin Seminar on High Energy Physics Problems: Relativistic Nuclear Physics and Quantum Chromodynamics (ISHEPP 2004)*

[22] Ermolov P F, Kokoulina E S, Kuraev E A, Kutov A Ya, Nikitin V A, Pankov A A, Roufanov I A and Zhidkov N K 2005 Study of multiparticle production by gluon dominance model. Part II. *17th International Baldin Seminar on High Energy Physics Problems: Relativistic Nuclear Physics and Quantum Chromodynamics (ISHEPP 2004)*

[23] Semenov S V, Troshin S M, Tyurin N E and Khrustalev O A 1975 Connection Between Elastic and Inelastic Processes at High-Energies. (In Russian) *Yad. Fiz.* **22** 792-800.

[24] Giovannini A 1979 QCD jets as Markov branching processes *Nucl.Phys.* **B**161 429-448.

[25] Barshay S 1989 Anamolous soft photons from a coherent hadronic phase in high-energy collisions. *Phys.Lett.* **B**227, N2, 279-284.

[26] White Paper http://theor.jinr.ru/twiki-cgi/view/NICA/NICAWhitePaper.

[27] Kokoulina E S 2002 Analysis of multiparticle dynamics in e^+e^- - annihilation into hadrons by two stage model. XXXII ISMD, Alushta, Ukraine *W.Sc. publ.* 340-343

[28] Kokoulina E S 2006 Gluon Dominance Model and e^+e^- annihilation, Proceedings of the 13th annual seminar "Nonlinear Phenomena in Complex Systems", Minsk, May 16-19, Joint Institute for Power and Nuclear Research-Sosny, NAS of Belarus **13** 73-82.

[29] Kuraev E A, Bakmaev S and Kokoulina E S 2011 Azimuthal correlation of gluon jets created in pp, $p\bar{p}$ and e^+e^- collisions, *Nucl.Phys.* **B**851 551-564.

[30] Didenko L A, Murzin V S and Sarycheva L I 1979 Leading and charged exchange in pi-p interactions at 40 GeV (talk). Proceedings, 16th International Cosmic Ray Conference, Vol.6 29-33.

[31] Kuraev E A, Kokoulina E S and Tomasi-Gustafsson E 2013 Hard light meson production in (anti)proton-hadron collisions and charge-exchange reactions. e-print arxiv.org hep/ph-1306.5169.

Pomeron in the $\mathcal{N}=4$ SYM at strong couplings

A V Kotikov[1], **L N Lipatov**[2]

[1] Bogoliubov Laboratory of Theoretical Physics, Joint Institute for Nuclear Research, 141980 Dubna, Russia

[2] Theoretical Physics Department, Petersburg Nuclear Physics Institute, 188300, Gatchina, Russia

E-mail: kotikov@theor.jinr.ru, lipatov@mail.desy.de

Abstract. We show the result for the BFKL Pomeron intercept at $\mathcal{N}=4$ Supersymmetric Yang-Mills model in the form of the inverse coupling expansion $j_0 = 2 - 2\lambda^{-1/2} - \lambda^{-1} + 1/4\,\lambda^{-3/2} + 2(1+3\zeta_3)\lambda^{-2} + O(\lambda^{-5/2})$, which has been calculated recently in [1] with the use of the AdS/CFT correspondence.

1. Introduction

The investigation of the high energy behavior of scattering amplitudes in the $\mathcal{N}=4$ Supersymmetric Yang-Mills (SYM) model [2]-[?] is important for our understanding of the Regge processes in QCD. Indeed, this conformal model can be considered as a simplified version of QCD, in which the next-to-leading order (NLO) corrections [8, 9] to the Balitsky-Fadin-Kuraev-Lipatov (BFKL) equation [10]-[14] are comparatively simple and numerically small. In the $\mathcal{N}=4$ SYM the equations for composite states of several reggeized gluons and for anomalous dimensions (AD) of quasi-partonic operators turn out to be integrable at the leading logarithmic approximation [15, 16, 17]. Further, the eigenvalue of the BFKL kernel for this model has the remarkable property of the maximal transcendentality [3]. This property gave a possibility to calculate the AD γ of the twist-2 Wilson operators in one [18, 19], two [3, 20], three [21], four [22, 23] and five [24] loops using the QCD results [25] and the asymptotic Bethe ansatz [26] improved with wrapping corrections [23] in an agreement with the BFKL predictions [2, 3].

On the other hand, due to the AdS/CFT-correspondence [28, 29, 30], in $\mathcal{N}=4$ SYM some physical quantities can be also computed at large couplings. In particular, for AD of the large spin operators Beisert, Eden and Staudacher constructed the integral equation [31] with the use the asymptotic Bethe-ansatz. This equation reproduced the known results at small coupling constants and is in a full agreement (see [32, 33, 34]) with large coupling predictions [35, 36, 37].

With the use of the BFKL equation in a diffusion approximation [2, 5], strong coupling results for AD [35, 36, 37] and the pomeron-graviton duality [38, 39] the Pomeron intercept was calculated at the leading order in the inverse coupling constant (see the Erratum[40] to the paper [21]). Similar results were obtained also in Ref. [41]. The Pomeron-graviton duality in the $\mathcal{N}=4$ SYM gives a possibility to construct the Pomeron interaction model as a generally covariant effective theory for the reggeized gravitons [42].

Below we present the strong coupling corrections to the Pomeron intercept $j_0 = 2 - \Delta$ in next orders. These corrections were obtained in Ref. [1] with the use of the recent calculations [43]-[47] of string energies.

2. BFKL equation at small coupling constant

The eigenvalue of the BFKL equation in $\mathcal{N}=4$ SYM model has the following perturbative expansion [2, 3] (see also Ref. [5])

$$j-1=\omega=\frac{\lambda}{4\pi^2}\left[\chi(\gamma_{BFKL})+\delta(\gamma_{BFKL})\frac{\lambda}{16\pi^2}\right],\quad \lambda=g^2N_c, \tag{1}$$

where λ is the t'Hooft coupling constant. The quantities χ and δ are functions of the conformal weights m and $\widetilde{m}$ of the principal series of unitary Möbius group representations, but for the conformal spin $n=m-\widetilde{m}=0$ they depend only on the BFKL anomalous dimension

$$\gamma_{BFKL}=\frac{m+\widetilde{m}}{2}=\frac{1}{2}+i\nu \tag{2}$$

and are presented below [2, 3]

$$\chi(\gamma) = 2\Psi(1)-\Psi(\gamma)-\Psi(1-\gamma), \tag{3}$$

$$\delta(\gamma) = \Psi''(\gamma)+\Psi''(1-\gamma)+6\zeta_3-2\zeta_2\chi(\gamma)-2\Phi(\gamma)-2\Phi(1-\gamma). \tag{4}$$

Here $\Psi(z)$ and $\Psi'(z)$, $\Psi''(z)$ are the Euler Ψ -function and its derivatives. The function $\Phi(\gamma)$ is defined as follows

$$\Phi(\gamma)=2\sum_{k=0}^{\infty}\frac{1}{k+\gamma}\,\beta'(k+1),\quad \beta'(z)=\frac{1}{4}\left[\Psi'\left(\frac{z+1}{2}\right)-\Psi'\left(\frac{z}{2}\right)\right]. \tag{5}$$

Due to the symmetry of ω to the substitution $\gamma_{BFKL}\to 1-\gamma_{BFKL}$ expression (1) is an even function of ν

$$\omega=\omega_0+\sum_{m=1}^{\infty}(-1)^m\,D_m\,\nu^{2m}, \tag{6}$$

where

$$\omega_0 = 4\ln 2\,\frac{\lambda}{4\pi^2}\left[1-\overline{c}_1\frac{\lambda}{16\pi^2}\right]+O(\lambda^3), \tag{7}$$

$$D_m = 2\left(2^{2m+1}-1\right)\zeta_{2m+1}\frac{\lambda}{4\pi^2}+\frac{\delta^{(2m)}(1/2)}{(2m)!}\,\frac{\lambda^2}{64\pi^4}+O(\lambda^3). \tag{8}$$

According to Ref. [3] we have

$$\overline{c}_1 = 2\zeta_2+\frac{1}{2\ln 2}\left(11\zeta_3-32\mathrm{Ls}_3\left(\frac{\pi}{2}\right)-14\pi\zeta_2\right)\approx 7.5812,\quad \mathrm{Ls}_3(x)=-\int_0^x\ln^2\left|2\sin\left(\frac{y}{2}\right)\right|dy. \tag{9}$$

Due to the Möbius invariance and hermicity of the BFKL hamiltonian in $\mathcal{N}=4$ SYM expansion (6) is valid also at large coupling constants. In the framework of the AdS/CFT correspondence the BFKL Pomeron is equivalent to the reggeized graviton [39]. In particular, in the strong coupling regime $\lambda\to\infty$

$$j_0 = 2-\Delta, \tag{10}$$

where the leading contribution $\Delta=2/\sqrt{\lambda}$ was calculated in Refs. [40, 41]. Below we find NLO terms in the strong coupling expansion of the Pomeron intercept.

3. AdS/CFT correspondence

Due to the energy-momentum conservation, the universal AD of the stress tensor $T_{\mu\nu}$ should be zero, i.e.,

$$\gamma(j=2)=0. \tag{11}$$

It is important, that the AD γ contributing to the DGLAP equation [48]-[52] does not coincide with γ_{BFKL} appearing in the BFKL equation. They are related as follows [8, 53, 54]

$$\gamma \;=\; \gamma_{BFKL}+\frac{\omega}{2} \;=\; \frac{j}{2}+i\nu\,, \tag{12}$$

where the additional contribution $\omega/2$ is responsible in particular for the cancelation of the singular terms $\sim 1/\gamma^3$ obtained from the NLO corrections (1) to the eigenvalue of the BFKL kernel [8]. Using above relations one obtains

$$\nu(j=2)=i\,. \tag{13}$$

As a result, from eq. (6) for the Pomeron trajectory we derive the following representation for the correction Δ (10) to the graviton spin 2

$$\Delta \;=\; \sum_{m=1}^{\infty} D_m. \tag{14}$$

According to (10) and (14), we have the following small-ν expansion for the eigenvalue of the BFKL kernel

$$j-2=\sum_{m=1}^{\infty} D_m\left(\left(-\nu^2\right)^m-1\right), \tag{15}$$

where ν^2 is related to γ according to eq. (12)

$$\nu^2=-\left(\frac{j}{2}-\gamma\right)^2. \tag{16}$$

On the other hand, due to the ADS/CFT correspondence the string energies E in dimensionless units are related to the AD γ of the twist-two operators as follows [29, 30][1]

$$E^2=(j+\Gamma)^2-4, \quad \Gamma=-2\gamma \tag{17}$$

and therefore we can obtain from (16) the relation between the parameter ν for the principal series of unitary representations of the Möbius group and the string energy E

$$\nu^2=-\left(\frac{E^2}{4}+1\right). \tag{18}$$

This expression for ν^2 can be inserted in the r.h.s. of Eq. (15) leading to the following expression for the Regge trajectory of the graviton in the anti-de-Sitter space

$$j-2=\sum_{m=1}^{\infty} D_m\left[\left(\frac{E^2}{4}+1\right)^m-1\right]. \tag{19}$$

[1] Note that our expression (17) for the string energy E differs from a definition, in which E is equal to the scaling dimension Δ_{sc}. But eq. (17) is correct, because it can be presented as $E^2=(\Delta_{sc}-2)^2-4$ and coincides with Eqs. (45) and (3.44) from Refs. [29] and [30], respectively.

4. Graviton Regge trajectory and Pomeron intercept

We assume, that eq. (19) is valid also at large j and large λ in the region $1 \ll j \ll \sqrt{\lambda}$, where the strong coupling calculations of energies were performed [43, 47]. These energies can be presented in the form [2]

$$\frac{E^2}{4} = \sqrt{\lambda}\frac{S}{2}\left[h_0(\lambda) + h_1(\lambda)\frac{S}{\sqrt{\lambda}} + h_2(\lambda)\frac{S^2}{\lambda}\right] + O\left(S^{7/2}\right), \tag{20}$$

where

$$h_i(\lambda) = a_{i0} + \frac{a_{i1}}{\sqrt{\lambda}} + \frac{a_{i2}}{\lambda} + \frac{a_{i3}}{\sqrt{\lambda^3}} + \frac{a_{i2}}{\lambda^2}. \tag{21}$$

The contribution $\sim \sqrt{S}$ can be extracted directly from the Basso result [44, 45] taking $J_{an} = 2$ according to [46]:

$$h_0(\lambda) = \frac{I_3(\sqrt{\lambda})}{I_2(\sqrt{\lambda})} + \frac{2}{\sqrt{\lambda}} = \frac{I_1(\sqrt{\lambda})}{I_2(\sqrt{\lambda})} - \frac{2}{\sqrt{\lambda}}, \tag{22}$$

where $I_k(\sqrt{\lambda})$ is the modified Bessel functions. It leads to the following values of coefficients a_{0i}

$$a_{00} = 1, \quad a_{01} = -\frac{1}{2}, \quad a_{02} = a_{03} = \frac{15}{8}, \quad a_{04} = \frac{135}{128} \tag{23}$$

The coefficients a_{10} and a_{20} come from considerations of the classical part of the folded spinning string corresponding to the twist-two operators (see, for example, [47])

$$a_{10} = \frac{3}{4}, \quad a_{20} = -\frac{3}{16}. \tag{24}$$

The one-loop coefficient a_{11} is found recently in the paper [46], considering different asymptotical regimes with taking into account the Basso result [44] (ζ_3 is the Euler ζ-function)

$$a_{11} = \frac{3}{16}\left(1 - \zeta_3\right), \tag{25}$$

Comparing the l.h.s. and r.h.s. of (19) at large j values gives us the coefficients D_m and Δ (see Appendix A in [1]).

5. Conclusion

We have shown the intercept of the BFKL pomeron at weak coupling regime and demonstrated an approach to obtain its values at strong couplings (for details, see Ref. [1]).

At $\lambda \to \infty$, the correction Δ for the Pomeron intercept $j_0 = 2 - \Delta$ has the form [3]

$$\Delta = \frac{2}{\lambda^{1/2}}\left[1 + \frac{1}{2\lambda^{1/2}} - \frac{1}{8\lambda} - \left(1 + 3\zeta_3\right)\frac{1}{\lambda^{3/2}} + \left(2a_{12} - \frac{145}{128} - \frac{9}{2}\zeta_3\right)\frac{1}{\lambda^2} + O\left(\frac{1}{\lambda^{5/2}}\right)\right]. \tag{26}$$

The fourth corrections in (26) contain unknown coefficient a_{12}, which will be obtained after the evaluation of spinning folded string on the two-loop level. Some estimations were given in Section 6 of [1] .

Acknowledgments

A.V.K. was supported in part by RFBR grant No. 13-02-01060-a. He thanks Eugene Levin and the Organizing Committee of International Moscow Phenomenology Workshop for invitation.

[2] Here we put $S = j - 2$, which in particular is related to the use of the angular momentum $J_{an} = 2$ in calculations of Refs [43, 47].

[3] Using a similar approach, the coefficients $\sim \lambda^{-1}$ and $\sim \lambda^{-3/2}$ were calculated also in the paper [55]. After correction of some errors, the results in [55] coincide with ours.

References

[1] Kotikov A V and Lipatov L N 2013 *Nucl. Phys.* B **874** 889
[2] Kotikov A V and Lipatov L N 2000 *Nucl. Phys.* B **582** 19
[3] Kotikov A V and Lipatov L N 2003 *Nucl. Phys.* B **661** 19
[4] Kotikov A V and Lipatov L N 2001 DGLAP and BFKL evolution equations in the N=4 supersymmetric gauge theory *Preprint* hep-ph/0112346.
[5] Fadin V F and Fiore R 2008 *Phys. Lett.* B **661** 139
[6] Fadin V F, Fiore R and Grabovsky A V 2010 *Nucl. Phys.* B **831** 248
[7] Balitsky I and Chirilli G A 2010 *Phys. Lett.* B **687** 204
[8] Fadin V F and Lipatov L N 1998 *Phys. Lett.* B **429** 127
[9] Camici G and M. Ciafaloni M *Phys. Lett.* B **430** 349
[10] Lipatov L N 1976 *Sov. J. Nucl. Phys.* **23** 338
[11] Fadin V F, Kuraev E A and Lipatov L N 1975 *Phys. Lett.* B **60** 50
[12] Fadin V F, Kuraev E A and Lipatov L N 1975 *Sov. Phys. JETP* **44** 443
[13] Fadin V F, Kuraev E A and Lipatov L N 1977 *Sov. Phys. JETP* **45** 199
[14] Balitsky I I and Lipatov L N 1978 *Sov. J. Nucl. Phys.* **28** 822
[15] Lipatov L N 1993 *Phys. Lett.* B **309** 394
[16] Lipatov L N 1993 High energy asymptotics of mult-color QCD and exactly solvable lattice models it Preprin hep-th/9311037
[17] Lipatov L N 1997 *Proc, Int. Conf. "Perspectives in Hadronic Physics" (Triest)* (Singapore: World Scientific) p. 562
[18] Lipatov L N 2001 *Nucl. Phys. Proc. Suppl.* A **99** 175
[19] Dolan F A and Osborn H 2002 *Nucl. Phys.* B **629** 3
[20] Kotikov A V, Lipatov L N and Velizhanin V N 2003 *Phys. Lett.* B **557** 114
[21] Kotikov A V, Lipatov L N, Onishchenko A I and Velizhanin V N 2004 *Phys. Lett.* B **595** 521
[22] Kotikov A V, Lipatov L N, Rej A, Staudacher M and Velizhanin V N 2007 *J. Stat. Mech.* **0710** P10003
[23] Bajnok Z, Janik R A and Lukowski T 2009 *Nucl. Phys.* B **816** 376
[24] Lukowski T, Rej A and Velizhanin V N 2010 *Nucl. Phys.* B **831** 105
[25] Moch S, Vermaseren J A M and Vogt A 2004 *Nucl. Phys.* B **688** 101
[26] Beisert N and Staudacher M 2005 *Nucl. Phys.* B **727** 1
[27] Staudacher M 2005 *JHEP* **0505** 054
[28] Maldacena J 1998 *Adv. Theor. Math. Phys.* **2** 231
[29] Gubser S S, Klebanov I R and Polyakov A M 1998 *Phys. Lett.* B **428** 105
[30] Witten E 1998 *Adv. Theor. Math. Phys.* **2** 253
[31] Beisert N, Eden B and Staudacher M 2007 *J. Stat. Mech.* **0701** P021
[32] Benna M K, Benvenuti S, Klebanov I R and Scardicchio A 2007 *Phys. Rev. Lett.* **98** 131603
[33] Kotikov A V and Lipatov L N 2007 *Nucl. Phys.* B **769** 217
[34] Basso B, Korchemsky G P and Kotanski J 2008 *Phys. Rev. Lett.* **100** 091601
[35] Gubser S S , Klebanov I R and Polyakov A M 2002 *Nucl. Phys.* B **636** 99
[36] Frolov S and Tseytlin A A 2002 *JHEP* **0206** 007
[37] Roiban R , Tirziu A and Tseytlin A A 2007 *JHEP* **0707** 056
[38] Polchinski J and Strassler M J 2002 *Phys. Rev. Lett.* **88** 031601
[39] Polchinski J and Strassler M J 2003 *JHEP* **0305** 012
[40] Kotikov A V, Lipatov L N, Onishchenko A I and Velizhanin V N 2006 *Phys. Lett.* B **632** 754
[41] Brower R C, Polchinski J, Strassler M J and Tan C I 2007 *JHEP* **0712** 005
[42] Lipatov L N 2013 *Phys. Part. Nucl.* **44** 391
[43] Gromov N, Serban D, Shenderovich I and Volin D 2011 *JHEP* **1108** 046
[44] Basso B 2011 An exact slope for AdS/CFT *Preprint* arXiv:1109.3154 [hep-th]
[45] Basso B 2912 Scaling dimensions at small spin in N=4 SYM theory *Preprint* arXiv:1205.0054 [hep-th]
[46] Gromov N and Valatka S 2012 *JHEP* **1203** 058
[47] Roiban R and Tseytlin A A 2011 *Nucl. Phys.* B **848** 251
[48] Gribov V N and Lipatov L N 1972 *Sov. J. Nucl. Phys.* **15** 438
[49] Gribov V N and Lipatov L N 1972 *Sov. J. Nucl. Phys.* **15** 675
[50] Lipatov L N 1975 *Sov. J. Nucl. Phys.* **20** 94
[51] Altarelli G and Parisi G 1977 *Nucl. Phys.* B **126** 298
[52] Dokshitzer Yu L 1977 *Sov. Phys. JETP* **46** 641
[53] Salam G P 1998 *JHEP* **9807** 019
[54] Salam G P 1999 *Acta Phys. Polon.* B **30** 3679
[55] Costa M S, Goncalves V and Penedones J 2012 *JHEP* **1212** 091

Average transverse momenta in hyperon production at p-p collider experiments

Olga Piskounova

P.N.Lebedev Physics Institute of Russian Academy of Science, Leninski prosp. 53, 119991 Moscow, Russia

E-mail: piskoun@sci.lebedev.ru

Abstract. The transverse momentum spectra of Λ^0 hyperon from LHC experiments (ALICE, ATLAS, CMS in the comparison with earlier experiments are considered from the point of view of Quark-Gluon String Model (QGSM) as examples of typical baryon spectra at very high energies. The LHC data at $\sqrt{s} = 0.9$ and 7 TeV and STAR data at $\sqrt{s} = 200$ GeV are fitted with the universal QGSM formula that includes the energy dependent slope as the main parameter. The dependence of average transverse momenta on $\sqrt{s}$ has been obtained with the help of this formula. The asymptotics of the energy dependence of $< p_t >$ shows the behavior $s^{0.055}$. This conclusion is very important for cosmic ray physics. It means that the long debated "knee" in the cosmic proton spectra at $E_p = (2,5-4)*10^{15}$ eV in laboratory system can not be considered any more as the result of dramatic changes in the dynamics of baryon hadroproduction. In the same time it may indicate the maximal energy of protons that can be produced in our Galaxy (or in other Calaxies).

1. Introduction

As you will see from the reference list to this paper, my study is following to the basical works of general authors of Quark-Gluon String Model: A.B.Kaidalov and K.A.Ter-Martirosyan [1]. Our collaboration began in early 1980th, while it was necessary to predict proton and pion spectra in whole kinematical regions (including the arrea of the beam particle fragmentation) that was caused by the problems in cosmic ray physics. Practically, only quark-gluon strings have been able to account correctly the energy conservation between different regions of hadron spectra. The entire corpus of the data on hadron spectra that have been measured at collider experiments those times was described and the cross sections for the production of heavy quark hadrons were predicted. In 90th, when it became clear that the quark structure of interacting particles can be revealed in the charge asymmetry of produced hadron spectra, we have studied all posible effects of leading/nonleading asymmetries in heavy-quark hadron spectra. This asymmetry should be caused by different quark contents of interacting and produced hadrons. The approach doesn't require the absolute values of differencial cross sections.

Only one suject of QGSM application left non-upgraded with new data, it is the description of transverse momentum spectra of produced particles. So, now I continue this activity on the new level of up-to-date collider investigation.

The transverse momentum spectra of Λ^0 hyperons that were measured by many hadron collider experiments: ISR [2], STAR [3], UA1 [4], UA5 [5], CDF [6] and experiments at LHC [7, 8, 9] show the form of distributions which is changing from lower energy to higher. As it was

Journal of Physics: Conference Series **607** (2015) 012012 doi:10.1088/1742-6596/607/1/012012

established in the previous publication [10], there is specific dependence of spectra on the type of colliding hadrons. Due to this difference, we have to confine ourself to the considering only the proton-proton collision results : ISR, STAR, ALICE, ATLAS and CMS in the wide range of $\sqrt{s}$: beginning from 53 GeV to LHC energy 7 TeV.

This range of energy has motivated me to investigate the behaviors of baryon spectra, because the proton spectrum in cosmic rays (CR) shows the "knee" at the very energy between Tevatron and LHC energies [11], see Fig.1. The change in the slope of proton spectrum may be of the astrophysical origin otherwise it means a substantial change in dynamics of particle production. We have to study the behaviors of baryon production at the collider experiments moving from one energy to another in order to conclude about whether or not the dramatic changes have place in hadroproduction processes at the "knee" energy and a little above.

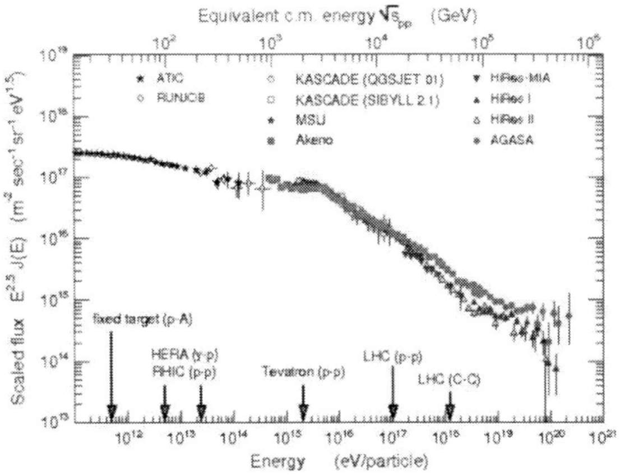

Figure 1. The cosmic proton spectrum at the energies between Tevatron and LHC.

The average transverse momenta of hyperons can show the specifical points in the characteristics of baryon production especially in the given range of energies. As it was learned in QGSM, the typical average transverse momentum at the production of certain hadrons is almost constant. It should slightly grow with energy due to the growing contributions from multipomeron exchanges that gives more fluctuations in transverse momenta.

2. Baryon transverse momentum distributions at LHC and QGSM description.

The recent data on Λ^0 hyperon distributions are obtained in the following LHC groups: ALICE [7] at 900 GeV, ATLAS [8] at 900 GeV and 7 TeV and CMS [9] at 900 GeV and 7 TeV. The lower energy experiment, which we can compare with the results of LHC groups, is STAR [3]($\sqrt{s} = 200GeV$).

If we fit the data with a simple exponential function e^{-B*p_t}, experiments, ATLAS and CMS, have presented the spectra with the similar slopes, see Fig.2. The different form of the distributions at low p_t's may be caused by specifics in the efficiency of the detecting procedure. It should be mentioned here that ATLAS spectra have systematically low efficiency at $p_t < 1$ GeV in comparison to the results of other LHC groups.

It was studied in early QGSM paper [12] that the transverse momentum spectra of hadrons after proton-proton collisions can be perfectly described with a bit more complicate dependence:

$$\frac{dN^H}{dp_t} = p_t * E\frac{d^3\sigma^H}{dx_F d^2p_t}\ p_t * e^{-B_0*(m_t - m_0)}, \tag{1}$$

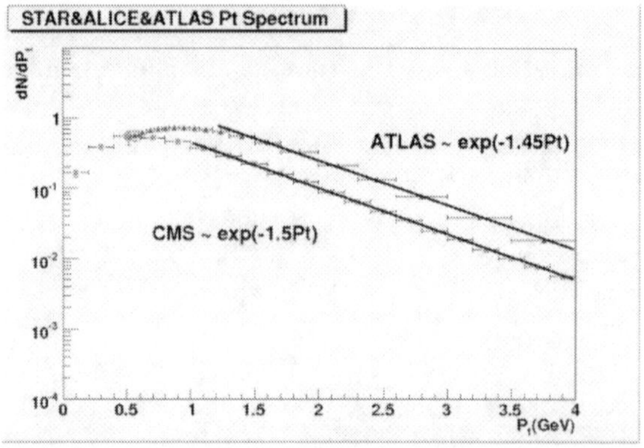

Figure 2. The comparison of ATLAS and CMS data at $\sqrt{s} = 7TeV$.

where m_0 is the mass of produced hadron, $m_t = \sqrt{p_t^2 + m_0^2}$ and B_0 used to bring the dependence on x_F, but in central region of rapidity this slope is constant. The slopes of the spectra of many types of hadrons (π, K, p and Λ^0) were estimated for the data of early proton-proton collision experiments of the energies that were available those times. The slopes of baryon spectra were approximately constant, $B_0 = 6.0$, which has been considered to be stable at all energies. Now we have to conclude that B_0 depends strongly on the energy. More, as it is seen from the spectra at LHC and RHIC, the value of m_0 is not equal nor to proton nor to hyperon masses.

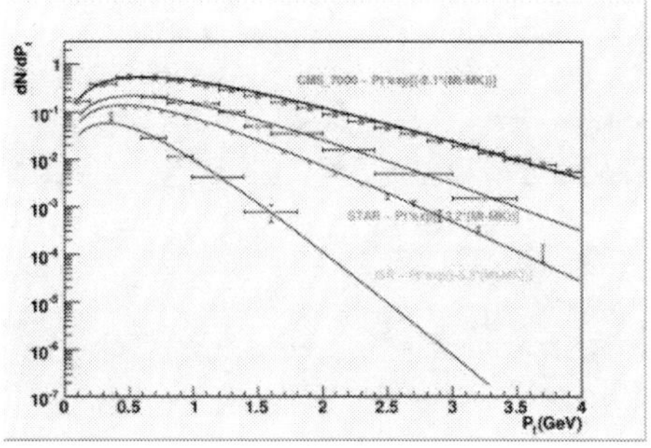

Figure 3. The description of STAR, ALICE and CMS data with the QGSM function.

The better description of hyperon spectra, which is shown in Fig.4, can be achieved with m_0 = 0,5 GeV that is kaon mass. This fact can be explained by the minimal quark-antiquark chain at hyperon production. In the short view, this chain is consisted of $K + \Lambda^0$ (only in the case of proton-proton collision) and the minimal transverse momentum of Λ, such a way, should be of order of kaon mass. In suggested fit, the values of slope parameter have to be equal for all LHC experiments of the same energy. It means that as soon as we estimate the slopes for various collider energies we are getting a chance to calculate (using the formula above) the energy dependence of average transverse momentum for the wide range of energy of proton-proton collisions.

3. Average transverse momenta at $\sqrt{s}$ = 53, 200 GeV, 900 GeV and 7 TeV

The resulting dependence of average p_t on the energy of interaction is shown in Fig.5. It seems reasonable to expect that the energy dependence of the average p_t value would somehow reproduce the multipomeronic behavior of high energy hadron interactions, which have been studied in QGSM within three decades, see [1, 13, 14, 15, 16]. The enhancing of the contributions from multipomeron components in the production diagrams has to cause the growing of fluctuations in transverse momenta of proton constituences that should result in the stable slight growing of the average p_t.

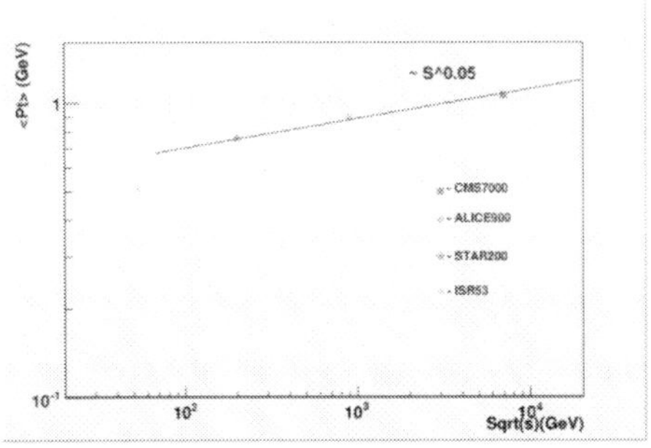

Figure 4. The average transverse momentum of hyperons vs. the energy of colliding protons.

We could only suggest that the fast growing of average p_t before STAR energy is due to valuable impact of baryon production from fragmentation region of beam proton. For the higher energy, the leading production of Λ^0 doesn't play an important role.

4. Conclusions

The review of results on transverse momentum distributions of hyperons that are produced in proton-proton collisions of up-to-date energies reveales a significant change in the slopes of spectra in the region of p_t = 0,3 - 4,0 GeV/c. The spectra of baryons are becoming harder and harder with the energy growth from RHIC ($\sqrt{s}$=200 GeV) to LHC (0,9 and 7 TeV). The detailed analysis of hyperon spectra, which is analogous to our early studies in frameworks of Quark-Gluon String Model, demonstrates the change of slopes from B_0 = 5,7 (ISR at 53 GeV) to B_0 = 2,1 (LHC at 7 TeV).

As a result, the average $< p_t >$ value is growing very fast up to approximately $\sqrt{s}$= 200 and then it goes with the asymptotics $s^{0.05}$. This behavior can not be considered as important change in the hadroproduction processes. Thus, it makes us conclude that processes that are taking place in baryon production at the energies of LHC are not anything unpredictable. This statement is very important for cosmic ray physics, where the "knee" (the change in the slope) at $E_{lab} \approx 3^* \ 10^{15}$ eV in cosmic proton spectra might have an origin in hadronic interactions. As we have discussed above, nothing unpredictable happens with baryon spectra up to $\sqrt{s}$ = 7 TeV that corresponds to $E_{lab} = 2,5 * 10^{16}$ eV. It means that the "knee" can be caused only due to astrophysical reasons. On the other hand, the "knee" may indicate the maximal energy of proton that are beeing produced in other Galaxies. But the idea of production of very high energy protons in space assumes a futher detailed investigations of the production dynamics of quark systems by means of QGSM.It seems very intriguing to observe whether or not the average transverse momenta of baryons become constant at futher energies of LHC.

Some brief analysis of growing-with-energy antiparticle-to-particle ratious of elementary particles that are measured in cosmic experiments was done in this QGSM technics as well. The growing ratios may be explained by the leading behavior of hadron production spectra that is inavitable result of "positiveness" of our Universe. The antibaryon/baryon production asymmetries, which are already measured in LHC experiments, are intended to be discussed in the upcoming publication.

5. Acknowledgements

I have to express my gratitude to the noncommercial organizations, Russian Foundation of Basic Research (grant RFBR- 13-02-06091) and Dmitry Zimin's "Dynasty" foundation, whoes financial supports made possible the First Kaidalov's Phenomenology Workshop. I hope as well that Kaidalov contributions into the phenomenology of QCD physics will give more and more results in the explanation of the contemporary experimental hadroproduction data.

References

[1] A.B.Kaidalov, Phys.Lett B**116**, 459(1982), Sov.J.Nucl.Phys.bf 45(1987)1450, A.B.Kaidalov and K.A.Ter-Martirosyan, Sov.J.Nucl.Phys.**39(1984)979; 40(1984)135.**
[2] **ISR Collaboration, D.Drijard** *et.al.*, **Z.Phys. C12, 217, (1982).**
[3] **STAR Collaboration, B.I. Abelev** *et.al.*, **Phys. Rev. C75, 064901,2007.**
[4] **UA1 Collaboration, G. Bocquet** *et.al.*, **Z. Phys. C366,441,1996.**
[5] **UA5 Collaboration, R.E. Ansorge** *et.al.*, **Nucl. Phys.B328,36,1989.**
[6] **CDF Collaboration, D. Acosta** *et.al.*, **Phys. Rev. D72, 052001,2005.**
[7] **ALICE Collaboration, EPJ C71,1594,2011, ArXiv:1012.3257.**
[8] **ATLAS Collaboration, Aad** *et.al.*, **PR D85,012001, ArXiv:1111.1297.**
[9] **CMS Collaboration, Khachatryan** *et.al.*, **JHEP 05,064, e-print -arXiv:1102.4282.**
[10] **O.I.Piskounova, e-print - arXiv:1209.6214 [hep-ph], 2012.**
[11] **V.B. Petkov, O.D.Lalakulich, G.M.Vereshkov, Phys.Atom.Nucl.66,523, 2003; Jorg R. Horandel, N.N.Kalmykov, A.V.Timokhin, Jou.of Phys.: Conference Series 47 (2006) 132.**
[12] **A.I. Veselov, O.I. Piskunova, K.A. Ter-Martirosian, Phys.Lett.B158, 175, 1985.**
[13] **Quark-Gluon String Model, A.B.Kaidalov and K.A.Ter-Martirosyan, Sov.J.Nucl.Phys.39, 1545, 1984, 40, 211, 1984, A.B.Kaidalov, Phys.Lett.B116, 459, 1982.**
[14] **A.B. Kaidalov and O.I. Piskunova Z. Phys. C30,145,1986, O.I. Piskounova Nucl.Phys.Proc.Suppl. 93, 144, 2001, e-print - hep-ph/0010263 and Phys.Atom.Nucl.70, 1107, 2007, e-Print: hep-ph/0604157.**
[15] **A.B.Kaidalov, O.I.Piskounova, Yad.Fiz. 43(1986)1545, O.I.Piskounova,Phys.Atom.Nucl.56, 1094, 1993.**
[16] **O.I.Piskounova, Phys.At.Nucl.64 392, 2001, e-Print: hep-ph/0001252.**
[17] **O.I.Piskounova, Phys.Atom.Nucl.66307,2003,e-Print: hep-ph/0202005 .**
[18] **O.I.Piskounova, Phys.Atom.Nucl.701107,2007,e-Print: hep-ph/0604157.**
[19] **O.I.Piskounova and N.V. Nikitin, Phys.Atom.Nucl.682124,2005,e-Print: hep-ph/0503006.**
[20] **O.I.Piskounova Sov.J.Nucl.Phys.51846,1990.**

Anisotropic pressure in the quark core of a strongly magnetized hybrid star

A A Isayev

[1]Kharkov Institute of Physics and Technology, Academicheskaya Street 1, Kharkov, 61108, Ukraine

E-mail: `isayev@kipt.kharkov.ua`

Abstract. The impact of a strong magnetic field, varying with the total baryon number density, on thermodynamic properties of strange quark matter (SQM) in the core of a magnetized hybrid star is considered at zero temperature within the framework of the Massachusetts Institute of Technology (MIT) bag model. It is clarified that the central magnetic field strength is bound from above by the value at which the derivative of the longitudinal pressure with respect to the baryon number density vanishes first somewhere in the quark core under varying the central field. Above this upper bound, the instability along the magnetic field is developed in magnetized SQM. The total energy density, longitudinal and transverse pressures are found as functions of the total baryon number density.

1. Introduction

The study of the QCD phase diagram under extreme conditions of temperature or density is continuing to be a hot research issue. The other factor which can significantly influence the structure of the QCD phase diagram is magnetic field. Strong magnetic fields of about $H \sim 10^{18}$ G (RHIC), or even an order of magnitude larger (LHC), are generated in noncentral high-energy heavy-ion collisions [1, 2]. The electric charge separation with respect to the reaction plane of colliding nuclei due to the the chiral magnetic effect [3] could be one of the observable imprints of strong magnetic fields generated in heavy-ion collisions. In the astrophysical context, for a special class of neutron stars, called magnetars, the magnetic field strength at the surface can reach the values of about 10^{14}-10^{15} G. Even stronger magnetic fields up to 10^{19} G may potentially occur in the inner cores of magnetars [4]. The possible imprint of such ultrastrong magnetic fields could be the large pulsar kick velocities as a result of asymmetric neutrino emission in direct Urca processes in the dense core of a magnetized neutron star [5]. The origin of strong magnetic fields of magnetars is still under discussion, and, among other possibilities, it is not excluded that this can be due to spontaneous ordering of hadron [6, 7, 8, 9], or quark [10] spins in the dense interior of a neutron star.

Thus, the study of thermodynamic properties of nuclear matter in a strong magnetic field is the problem of a considerable interest. In particular, the pressure anisotropy, exhibited in the difference between the longitudinal and transverse (along and perpendicular to the magnetic field) pressures, becomes important for strongly magnetized matter [11, 12, 13, 14, 15]. In this study, we consider strongly magnetized SQM, composed of deconfined up, down and strange quarks, within the framework of the MIT bag model [16, 17]. The relevant astrophysical object

Journal of Physics: Conference Series **607** (2015) 012013 doi:10.1088/1742-6596/607/1/012013

is a hybrid star whose quark core is surrounded with the hadronic crust. It was clarified earlier that, if the uniform magnetic field exceeds some critical value, the longitudinal pressure becomes negative resulting in the appearance of the longitudinal instability in SQM [14]. The value of the corresponding critical field represents, in fact, the upper bound on the magnetic field in the quark core of a hybrid star when spatial nonuniformity in the field distribution is disregarded. In this research, we would like to extend the previous consideration [14] to the case of the spatially nonuniform magnetic field distribution.

2. Numerical results and discussion

For the details of the formalism one can address to Ref. [14]. We use the simplified variant of the MIT bag model, in which quarks are considered as free fermions moving inside a finite region of space called a "bag". The effects of the quark confinement are implemented by introducing the bag pressure B. Note that SQM can represent the ground state of matter, or can be metastable. In order to be absolutely stable, the energy per baryon of magnetized SQM should be less than that of the most stable ^{56}Fe nucleus under the zero external pressure and temperature. Further we will be interested in the astrophysical scenario, in which SQM forms the core of a hybrid star and, hence, is metastable at zero pressure. The gravitational pressure from the outer hadronic layers stabilizes quark matter in the core.

In the MIT bag model, the total energy density E, the longitudinal p_l and transverse p_t pressures in magnetized quark matter read

$$E = \Omega + \sum_i \mu_i \varrho_i + \frac{H^2}{8\pi} + B, \tag{1}$$

$$p_l = -\Omega - \frac{H^2}{8\pi} - B, \tag{2}$$

$$p_t = -\Omega - HM + \frac{H^2}{8\pi} - B, \tag{3}$$

where $M = -\frac{\partial\Omega}{\partial H}$ is the total magnetization, $\Omega = \sum_i \Omega_i$, Ω_i and $\varrho_i = -\frac{\partial\Omega_i}{\partial\mu_i}$ are the thermodynamic potential and number density, respectively, for fermions of ith species with the chemical potential μ_i, including u, d, s quarks plus electrons to ensure charge neutrality and beta equilibrium with respect to the weak processes occurring in the quark core of a hybrid star.

In the previous research [14], the impact of the uniform magnetic field on thermodynamic properties of SQM was considered. In a more realistic study, it is necessary to take into account that the magnetic field varies from the core to the surface of a star. Following Ref. [4], we will model this change by the dependence of the magnetic field on the baryon density of the form

$$H(\varrho_B) = H_s + H_{cen}\left(1 - e^{-\beta\left(\frac{\varrho_B}{\varrho_0}\right)^\gamma}\right). \tag{4}$$

Here H_{cen} and H_s are the magnetic field strengths in the center (assuming that the central baryon density is essentially larger than the nuclear saturation density $\varrho_0 = 0.16$ fm^{-3}) and at the surface of a star, respectively; β and γ are the model parameters. We set the surface field $H_s = 10^{15}$ G, and the central density is chosen to be $\varrho_{cen} = 7\varrho_0$. In Eq. (4), we use the model parameter sets $\beta = 0.02, \gamma = 3$ from Ref. [18] (slow varying magnetic field), and $\beta = 0.001, \gamma = 6$ from Ref. [15] (fast varying magnetic field). In the subsequent calculations, the same quark current masses are employed as in the studies [14, 19], and we adopt $B = 76$ MeV/fm^3, which is slightly larger than the upper bound $B_u \simeq 75$ MeV/fm^3 from the absolute stability window.

Further we will assume that the quark phase appears at the total baryon density $\varrho_B \sim 3\varrho_0$ [20], and, hence, the baryon density for magnetized SQM changes in the range $3\varrho_0 \lesssim \varrho_B \leq 7\varrho_0$. The

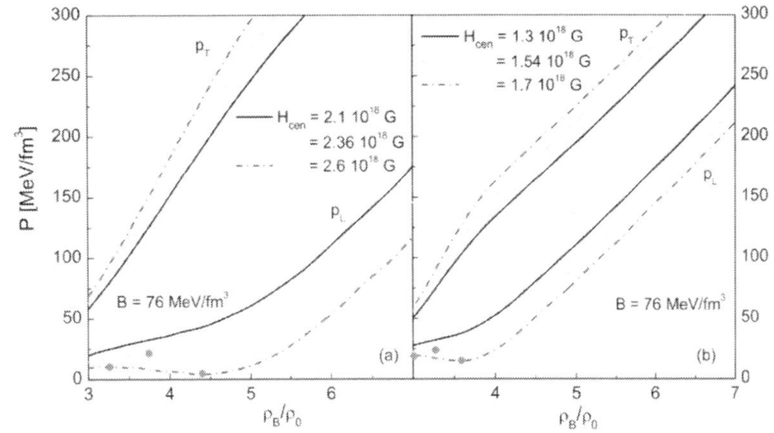

Figure 1. Transverse p_t (three upper curves) and longitudinal p_l (three lower curves) pressures in magnetized SQM as functions of the total baryon number density, corresponding to Eq. (4) with: (a) $\beta = 0.02, \gamma = 3$ and (b) $\beta = 0.001, \gamma = 6$ for $H_s = 10^{15}$ G and variable central field H_{cen}. The full dots correspond to the points where $p_l{}'(\varrho_B) = 0$.

main strategy in the further calculations of the anisotropic pressure is as follows. We vary the central magnetic field strength H_{cen} in Eq. (4), and like to determine at which baryon density within the above range the onset of the longitudinal instability occurs. Fig. 1 shows the transverse p_t and longitudinal p_l pressures as functions of the total baryon number density for the above parametrizations of the magnetic field strength. The general tendency is that, under increasing the central field H_{cen}, the transverse pressure p_t increases while the longitudinal pressure p_l decreases. Also, the transverse pressure always increases with the total baryon density ϱ_B, while the dependence of the longitudinal pressure p_l on ϱ_B can be different. Let us consider first the case of the slow varying magnetic field with $\beta = 0.02, \gamma = 3$ in Eq. (4) (Fig. 1a). Under increasing the central field H_{cen}, the longitudinal pressure, at first, remains to be increasing function of the baryon density with $p_l{}'(\varrho_B) > 0$. However, under further increasing the central field, the curve $p_l(\varrho_B)$ is bending down in its middle part. At $H_{cen} \approx 2.36 \cdot 10^{18}$ G, the derivative $p_l{}'(\varrho_B)$ vanishes at $\varrho_B \approx 3.74\varrho_0$ (the corresponding point on the curve is marked by the full red dot) while remaining positive for other baryon densities from the interval under consideration. Under further increasing the central magnetic field, there appears the part of the curve characterized by the negative derivative $p_l{}'(\varrho_B) < 0$, contrary to the stability constraint $p_l{}'(\varrho_B) > 0$. Hence, the corresponding states of magnetized SQM are unstable and the instability is developed along the magnetic field direction. The onset of instability corresponds to $H_{cen} \approx 2.36 \cdot 10^{18}$ G, at which the derivative $p_l{}'(\varrho_B)$ vanishes first. This value represents the upper bound on the central magnetic field strength in the quark core of a hybrid star in the case of the slow varying magnetic field with $\beta = 0.02, \gamma = 3$ in Eq. (4). For the fast varying field with $\beta = 0.001, \gamma = 6$, the upper bound on the central magnetic field strength is somewhat smaller, $H_{cen} \approx 1.54 \cdot 10^{18}$ G, for which the derivative $p_l{}'(\varrho_B)$ vanishes at $\varrho_B \approx 3.27\varrho_0$ (cf. Fig. 1b).

Fig. 2 shows the energy density E of the system and its matter part $E_m \equiv E - E_f$ ($E_f = \frac{H^2}{8\pi}$ being the magnetic field energy density) as functions of the total baryon number density for the above parametrizations of the magnetic field strength. The curves for the matter part are practically indistinguishable for the different values of the central magnetic field, used in calculations with each parameter set, and look almost as one curve. This figure allows to estimate the relative role of the matter E_m and magnetic field E_f contributions to the total energy density

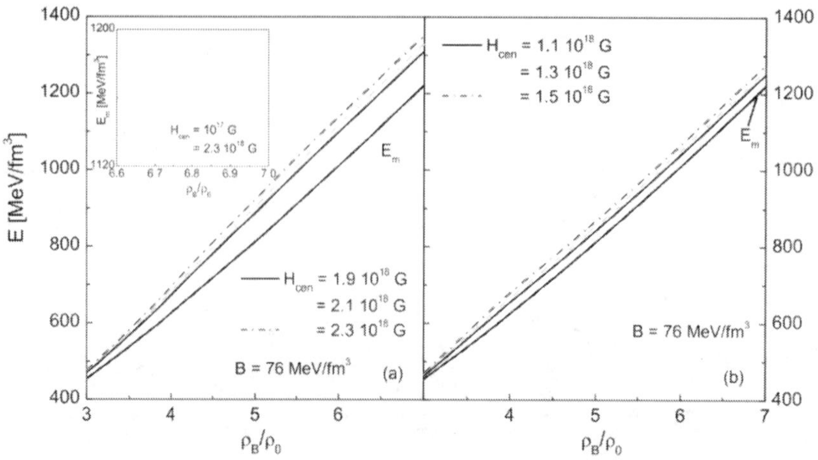

Figure 2. The energy density E (upper curves apart from the most lower curve) and the matter energy density E_m (the most lower curve) of magnetized strange quark matter vs. the total baryon number density for the same parametrizations of the magnetic field strength as in Fig. 1 and variable central field H_{cen}. The insert in the left panel shows the matter energy density E_m in the high baryon density region.

E. It is seen that the matter part dominates over the field part at such baryon number densities and magnetic field strengths H_s, H_{cen}. Also, the insert in the left panel shows the matter part E_m as a function of the baryon density for two differing by an order of magnitude values of the central field H_{cen} in order to demonstrate the effect of Landau diamagnetism. It is seen that the matter part is smaller in the case of the stronger central magnetic field. Nevertheless, the overall effect of the stronger central field is to increase the total energy of the system because of the pure magnetic field contribution.

In strong magnetic fields beyond some threshold value $H > H_{th}$, with $10^{17} < H_{th} \lesssim 10^{18}$ G, the total pressure in magnetized strange quark matter becomes essentially anisotropic [14]. Hence, the equation of state (EoS) of the system becomes also highly anisotropic. Fig. 3, showing the dependence of the transverse p_t (upper row) and longitudinal p_l (lower row) pressures on the energy density E of magnetized strange quark matter, explicitly demonstrates this moment. Because in all cases we choose the central field smaller than the corresponding upper bound, all quantities p_t, p_l and E are the increasing functions of the baryon density, and, hence, $p_t(E)$ and $p_l(E)$ are also the increasing functions.

In summary, we have considered the impact of varying with the total baryon number density magnetic field on thermodynamic properties of SQM at zero temperature under conditions relevant to the cores of strongly magnetized hybrid stars. The total energy density E of magnetized SQM, transverse p_t and longitudinal p_l pressures have been calculated as functions of the total baryon number density. Also, the highly anisotropic EoS has been determined in the form of $p_t(E)$ and $p_l(E)$ dependences. It has been clarified that the central magnetic field in the core of a hybrid star is bound from above by the value, at which the derivative of the longitudinal pressure $p_l{}'(\varrho_B)$ vanishes first under varying the central field. Above this upper bound, the instability along the magnetic field direction is developed in magnetized SQM.

In this study, we have considered the thermodynamic properties of strongly magnetized SQM at zero temperature. It would be of interest also to extend this research to finite temperatures [12, 21], when new interesting effects could occur, such as the unusual behavior of the entropy of a spin polarized state [22, 23, 24].

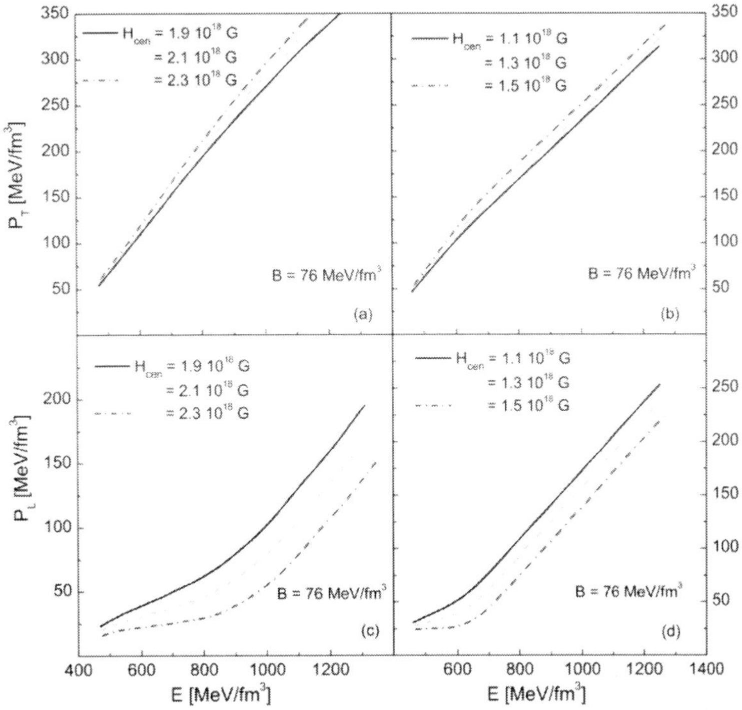

Figure 3. The transverse p_t (panels (a), (b)) and longitudinal p_l (panels (c), (d)) pressures in magnetized strange quark matter as functions of the total energy density E for the same parametrizations of the magnetic field strength as in Fig. 1 and variable central field H_{cen}.

References

[1] Skokov V, Illarionov A and Toneev V 2009 *Int. J. Mod. Phys.* A **24** 5925
[2] Deng W-T, and Huang X-G 2012 *Phys. Rev.* C **85** 044907
[3] Kharzeev D E, McLerran L D , Warringa H J 2008 *Nucl. Phys.* A **803** 227
[4] Bandyopadhyay D, Chakrabarty S and Pal S 1997 *Phys. Rev. Lett.* **79** 2176
[5] Adhya S P, Roy P K and Dutt-Mazumder A K 2014 *J. Phys.* G **41** 025201
[6] Isayev A A and Yang J 2004 *Phys. Rev.* C **69** 025801
[7] Isayev A A 2006 *Phys. Rev.* C **74** 057301
[8] Isayev A A and Yang J 2006 *Progress in Ferromagnetism Research*, ed V N Murray (New York: Nova Science Publishers) p 325
[9] Isayev A A and Yang J 2009 *Phys. Rev.* C **80** 065801
[10] Tatsumi T 2000 *Phys. Lett.* B **489** 280
[11] Khalilov V R 2002 *Phys. Rev.* D **65** 056001
[12] Isayev A A and Yang J 2011 *Phys. Rev.* C **84** 065802
[13] Isayev A A and Yang J 2012 *Phys. Lett.* B **707** 163
[14] Isayev A A and Yang J 2013 *J. Phys.* G **40** 035105
[15] Sinha M, Mukhopadhyay B and Sedrakian A 2013 *Nucl. Phys.* A **898** 43
[16] Farhi E and Jaffe R L 1984 *Phys. Rev.* D **30** 2379
[17] Chodos A et al. 1974 *Phys. Rev.* D **9** 3471
[18] Rabhi A, Pais H, Panda P K, and Providência C 2009 *J. Phys.* G **36** 115204
[19] Huang X-G, Huang M, Rischke D H, and Sedrakian A 2010 *Phys. Rev.* D **81** 045015
[20] Weber F 2005 *Prog. Part. Nucl. Phys.* **54** 193
[21] Dexheimer V, Menezes D P and Strickland M 2014 *J. Phys.* G **41** 015203
[22] Isayev A A 2005 *Phys. Rev.* C **72** 014313
[23] Isayev A A 2007 *Phys. Rev.* C **76** 047305
[24] Isayev A A and Yang J 2010 *J. Korean Astron. Soc.* **43** 161

Overview of Higgs boson results from ATLAS experiment at the LHC

I.I. Tsukerman on behalf of the ATLAS Collaboration
ITEP, Moscow, 117218, Russia
E-mail: ilya.tsukerman@cern.ch

Abstract. One year ago, a new particle with the mass of 125 GeV and properties expected for the Standard Model Higgs boson was discovered by ATLAS and CMS experiments at the Large Hadron Collider at CERN in proton-proton collisions at center-of-mass energies $\sqrt{s}$ =7 TeV and 8 TeV. Since that time more data at 8 TeV were accumulated and analyzed and a progress was reached in the measurement of the particle properties. In this note an overview of Higgs boson results obtained recently at ATLAS experiment is given. Complete dataset which corresponds to 4.7 fb^{-1} (20.5 fb^{-1}) of the data obtained at $\sqrt{s}$ =7 TeV (8 TeV) collisions, respectively, is used. The following four signal decay channels are considered: $H \to \gamma\gamma$, $H \to ZZ^* \to 4\ell$, $H \to WW^* \to \ell\ell\nu\nu$ and $H \to b\bar{b}$. These channels together with $H \to \tau\tau$ channel are combined to extract Higgs boson mass, strength, coupling constants, spin and parity. All experimental results are compatible with the Standard Model predictions.

1. Introduction

The search for the Standard Model (SM) Higgs boson (H) at the Large Hadron Collider (LHC) is known to be complicated task as expected signal production cross section is σ_H =O(10) pb while background rates are much higher. Fig.1(a) shows how predicted σ_H depends on its mass at different center-of-mass pp energies, $\sqrt{s}$ =7 TeV, 8 TeV and 14 TeV [1] which correspond to the LHC conditions in 2011, 2012 and 2015 year, respectively. It is about 15 pb at m_H =125 GeV at $\sqrt{s}$ =7 TeV being 1.3 times higher at $\sqrt{s}$ =8 TeV. At larger m_H the cross section increases with $\sqrt{s}$ stronger. There are four main mechanisms of H production at LHC energies: gluon-gluon fusion (ggF) via heavy-quark triangular loop, vector-boson fusion (VBF) where Higgs boson is accompanied by two energetic jets going at small polar angles, associated production with one vector boson (VH, i.e. together with W or Z) and top-antitop fusion. The expected σ_H in these mechanisms is shown in Fig.1(b) as function of m_H at $\sqrt{s}$=8 TeV together with theoretical uncertainties [1]. The σ_H was calculated in NNLO approximation (except $t\bar{t}H$ process for which NLO approach is used). It is seen that the ggF mechanism dominates while rates via VBF (VH, $t\bar{t}H$) mechanisms are ten (a few dozens, hundred) times smaller, respectively.

Expected branching ratios (BR) of the experimentally favourable decay modes multiplied by σ_H at 8 TeV are shown in Fig.2(a). They are as large as 1.5 pb at m_H =125 GeV for $H \to \tau\tau$ and $H \to WW^* \to \ell\nu q\bar{q}$ channels. However, signal-to-background (S/B) ratios for these channels are small and mass reconstruction is problematic. $H \to WW^* \to \ell\nu\ell\nu$ channel has lower cross section, 0.25 pb, but one can reach sizeable S/B ratio doing kinematical selections. Again, mass reconstruction looks non-trivial. $H \to \gamma\gamma$ channel has very good and simple signature. Hovewer, $\sigma_H \times BR(H \to \gamma\gamma)$ is less than 0.05 pb, and background conditions are very hard. The most clean decay mode, where both signature and S/B ratio are perfect is $H \to ZZ^{\star} \to 4\ell$. But $\sigma_H \times BR(H \to 4\ell)$ is only a few pb due to small leptonic branching ratio of the Z boson. Despite these difficulties, both ATLAS [2] and CMS [3] Collaborations

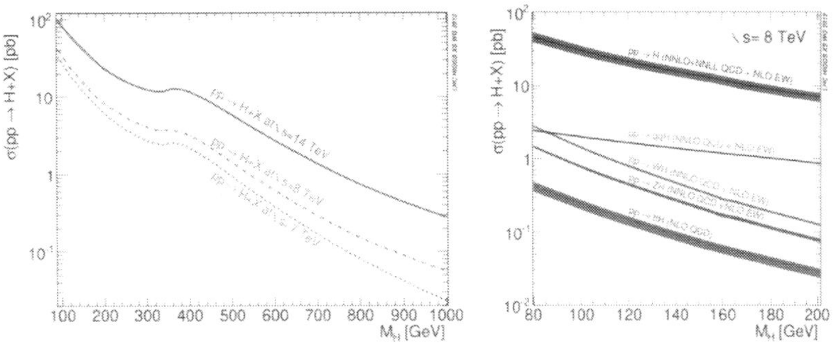

(a) (b)

Figure 1. (a) Predicted total σ_H at $\sqrt{s}$ =7, 8 and 14 TeV as function of m_H. (b) Predicted total σ_H at $\sqrt{s}$ =8 TeV together with separate contributions from different production mechanisms as function of m_H.

at the LHC were able to discover one year ago a new particle with properties of SM-like Higgs boson combining $H \to \gamma\gamma$, $H \to ZZ^{\star} \to 4\ell$ and $H \to WW^{\star} \to \ell\ell\nu\nu$ decay channels. Their studies were based on complete data taken at 7 TeV ($\approx$5 fb^{-1}) and on one third of 8-TeV dataset ($\approx$6 fb^{-1}). Currently, preliminary analyses of full 8 TeV data are advanced and corresponding results are started to appear. Summary presented in this note is based on complete 7-TeV and 8-TeV ATLAS datasets. The note is organized as follows. The ATLAS detector is briefly described in Section 2. In Section 3, the most interesting decay modes of the SM Higgs boson, namely $H \to \gamma\gamma$, $H \to ZZ^{\star} \to 4\ell$, $H \to WW^{\star} \to \ell\ell\nu\nu$ and $H \to b\bar{b}$ are considered; results of statistical combination of these channels together with $H \to \tau\tau$[1] are presented in Section 4. This combination includes particle mass, signal strength, coupling constants, spin and parity.

2. The ATLAS detector and its modeling

The ATLAS detector [4] is a multipurpose particle physics apparatus with forward-backward symmetric cylindrical geometry. The inner tracking detector is surrounded by a thin superconducting solenoid which provides a 2 T magnetic field, and by electromagnetic (EM) and hadronic calorimetry. The calorimeters are divided into a central barrel and end-cap regions on either end of the detector covering the range $|\eta| \leq 4.9$.[2] In the central region $|\eta| \leq 1.8$, the EM calorimeter is preceded by a presampler detector to correct for upstream energy losses. The muon spectrometer surrounds the calorimeters and consists of three air-core superconducting magnets providing a toroidal field, a system of precision tracking chambers, and fast detectors for triggering. The combination of all these systems provides charged particle measurements together with efficient and precise lepton and photon measurements in the range $|\eta| \leq 2.5$.

[1] For this particular channel, one-year-old results are used in the combination.

[2] ATLAS uses a right-handed coordinate system with the origin at the nominal interaction point (IP) in the centre of the detector, and the z-axis along the beam line. The x-axis points from the IP to the centre of the LHC ring, and the y-axis points upwards. Cylindrical coordinates (r, θ) are used in the transverse plane, θ being the azimuthal angle around the beam line. Observables labelled "transverse" are projected into the $x - y$ plane. The pseudorapidity is defined in terms of the polar angle θ, as $\eta = \ln\tan(\theta/2)$.

Jets and missing transverse energy, E_T^{miss}, are reconstructed using energy deposits over the full coverage of the calorimeters. To develop analysis procedures, to calculate signal acceptance and to evaluate contribution for some background processes, Monte-Carlo (MC) modeling is used. All simulated signal and background datasets are passed through GEANT4 [5] full simulation chain of ATLAS set-up and algorithms used also in real data recontruction. Multiple interactions in the same bunch crossing[3] including detector effects are also modeled. In many cases backgrounds are directly **measured** experimentally in the so-called control regions, kinematically orthogonal to signal regions. Corresponding normalization factors are determined from simulation.

3. $H \to \gamma\gamma$, $H \to ZZ^\star \to 4\ell$, $H \to WW^\star \to \ell\ell\nu\nu$ and $H \to b\bar{b}$ decay channels

3.1. $H \to \gamma\gamma$ *[6], [7], [8]*

The expected cross section of $H \to \gamma\gamma$ production and decay is 30–50 fb in the region m_H =110–150 GeV. At higher Higgs boson masses it falls rapidly mainly due to strong decrease of the BR($H \to \gamma\gamma$). The signature is two high-p_T isolated photons with invariant mass $m_{\gamma\gamma} = m_H$. Simulations show that the ATLAS EM calorimeter allows to reach invariant-mass resolution $FWHM \approx 4$ GeV in the region of interest. Irreducible background consists of continuum production of photon pairs via $q\bar{q} + gg \to \gamma\gamma$ mechanisms. These processes have about hundred times larger rates than expected for the signal. Reducible backgrounds include mostly γjet- and $jetjet$-production when one or two jets are misidentified as photons in the detector. These processes have many orders of magnitude higher cross sections than the continuum diphoton production. Thanks to the EM calorimeter with a preshower, one can reach required jet rejection at the level of 10^4. Some 13K events survive all the selections while expected signal in the mass window around 125 GeV is $\approx$ 400. To increase discovery potential, the events are further subdivided into fourteen independent categories having different expected $m_{\gamma\gamma}$ resolution and S/B ratio. The $m_{\gamma\gamma}$-distribution after corresponding reweighting is given in Fig.2(b) together with the spectrum after background subtraction. Excess of events is seen around 126 GeV. The probability that background fluctuation causes at least observed excess of events, called p_0, is shown in Fig.3(a) as function of m_H. A clear signal at around 126.5 GeV with 7.4σ local significance is seen. The measured mass of the new particle is 126.8±0.2 (stat)±0.7 (syst) GeV, where systematic error is mainly due to photon energy scale uncertainty. Taking into account known Landau-Yang theorem, this particle cannot have spin equal to one. The measured signal strengths in units of the SM for VH, VBF and ggF+$t\bar{t}H$ couplings separately, are the following: $\mu_{ggF+t\bar{t}H} \times B/B_{SM}$ =$1.6^{+0.3}_{-0.3}$(stat)$^{+0.3}_{-0.2}$(syst); $\mu_{VBF} \times B/B_{SM}$ =$1.7^{+0.8}_{-0.8}$(stat)$^{+0.5}_{-0.4}$(syst) and $\mu_{VH} \times B/B_{SM}$ =$1.8^{+1.5}_{-1.3}$(stat)$^{+0.3}_{-0.3}$(syst). Combining them one obtains the value μ =1.65±0.24 (stat)$^{+0.25}_{-0.18}$ (syst) which is 2.3σ away from the SM prediction of one.

Using kinematical distributions in the so-called Collins-Soper frame one can exclude a possibility of spin-2 of newly observed particle [7]. The obtained log-likelihood ratio of spin-0 and spin-2 hypotheses is given in Fig.3(b) as function of $f_{q\bar{q}} = \sigma_H^{q\bar{q}}/\sigma_H^{total}$. One can see that the spin-2 hypothesis[4] is excluded at 99% CL assuming that all signal events are produced through gg-fusion. This number is reduced to 95% if 25% admixture of $q\bar{q}$ events is supposed.

In Ref.[8] distributions on kinematic variables in di-photon signal candidates are analyzed and compared with theoretical models. No big deviation from the SM prediction is observed.

3.2. $H \to ZZ^\star \to 4\ell$ *[9]*

$H \to ZZ^\star \to 4\ell$ decay mode is promising when searching for the SM Higgs boson in a wide mass range from 110 GeV to 600 GeV. The $\sigma_H \times BR(H \to 4\ell)$ is 3 fb at m_H =125 GeV for $\sqrt{s}$=8 TeV. Signal signature is two pairs of isolated, high p_T opposite-sign leptons (electrons or muons);

[3] In average, ten (twenty) interactions occurred each 50 ns in the same bunch crossing at $\sqrt{s}$=7 TeV (8 TeV).
[4] Graviton-like model is used to simulate the spin-2 case.

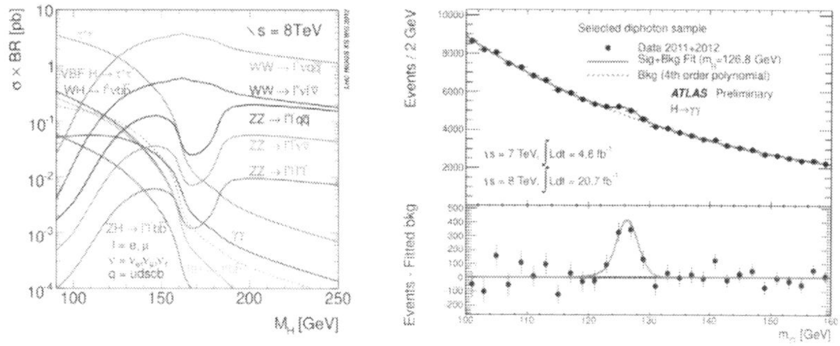

(a) (b)

Figure 2. (a) $\sigma_H \times BR(H \to X)$ at $\sqrt{s}$ =8 TeV for the most promising decay channels X as function of m_H. (b) $m_{\gamma\gamma}$-distribution for the combined 7-TeV and 8-TeV data samples in $H \to \gamma\gamma$ searches. The bottom inset displays the residuals of the data with respect to the fitted background component.

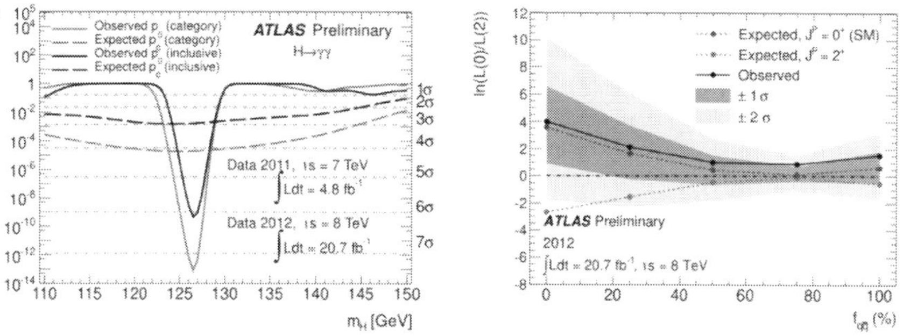

(a) (b)

Figure 3. Results of the $H \to \gamma\gamma$ study. (a) The observed local p_0 value as function of m_H for the combination of 7-TeV and 8-TeV data for the inclusive case (black) and for the analysis using categories (red). (b) Median expected values of the test statistics $ln(L(0)/L(2))$ for the spin-0 (dashed blue/upper line and corresponding 68% and 95% CL bands) and spin-2 (dashed red/lower line) hypotheses, as well as the observed values (solid line) as function of the $f_{q\bar{q}}$.

at least one pair should have $m_{\ell\ell} = m_Z$. Irreducible backgrounds come from $q\bar{q} + gg \to ZZ^\star$ processes which have comparable cross sections with the expected signal at 125 GeV. Smaller reducible backgrounds are due to $t\bar{t}$ and Drell-Yan production. ATLAS detector has excellent four-lepton invariant-mass resolution, namely 1.6 GeV (1.9 GeV, 2.4 GeV) for the 4μ ($2\mu 2$e, 4e) cases, respectively. The measured invariant-mass distribution, $m_{4\ell}$, after combination of all the lepton cases is shown in Fig.4(a). Clear maximum above background is seen in the region around 125 GeV; 32 events are observed in the mass window 125±5 GeV with the estimated background 11.1±1.0 events. The data/MC ratio is compatible with unity in the sidebands. The peak at $m_{4\ell} = m_Z$ is due to rare Z boson decay to four leptons. The p_0 probability is shown in Fig.4(b) as function of m_H in the region 110–180 GeV. The local significance reaches 6.6σ at 124 GeV and no other big fluctuations are observed in the mentioned mass range. The reconstructed mass of the new particle is 124.3±0.6 (stat)±0.4 (syst) GeV. The measured signal strength, μ =1.7±0.4, is compatible with the SM prediction of one at 2σ level. In addition, SM-like Higgs boson is excluded at 95% CL with the mass below 650 GeV (350 GeV) in the ggF (VBF) mechanisms alone [9]. Despite signal statistics is small, the excellent S/B ratio allows to study such properties of the signal as spin and parity using angular distributions of the newly observed particle's decay products. Six spin-parity hypotheses are tested: 0^+ (as predicted by the SM), 0^-, 1^+, 1^-, 2^+ and 2^-. The events passed all kinematical selections which have 115 GeV$\leq m_{4\ell} \leq$130 GeV are used in this study. Two analyses were performed. The first one is based on Boosted Decision Tree (BDT) and the second one is based on matrix element likelihood ratio (J^P-MELA). Log-likelihood ratios are presented in Table 1 for all the spin-parity cases as

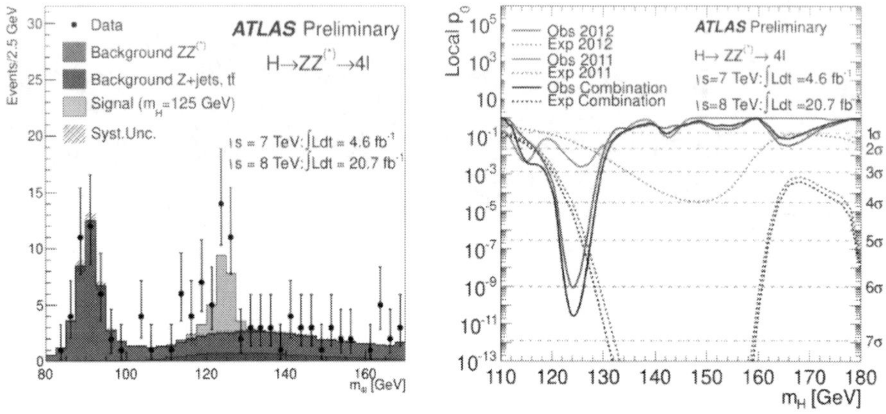

(a) (b)

Figure 4. Results of the $H \to ZZ^{\star} \to 4\ell$ study. (a) $m_{4\ell}$ distributions for the selected candidates compared to the background expectation in the mass range 80-170 GeV. The signal expectation for the m_H =125 GeV hypothesis is also shown. (b) The observed local p_0-value for the combination of the 2011 and 2012 data sets (solid black line); the $\sqrt{(s)}$ =7 TeV and 8 TeV data results are shown in solid lines (blue and red, respectively).

well as for the BDT-based and MELA-based analyses separately. One can see that the 0^- and 1^+ hypotheses are strongly disfavoured with respect to the 0^+ hypothesis. However, exclusion of the 2^+ case is less effective. The 0^+ and 2^+ hypotheses are also compared assuming different $f_{q\bar{q}}$ fraction [9].

3.3. $H \to WW^{\star} \to \ell\ell\nu\nu$ decay channel [10], [11], [12], [13]

$H \to WW^{\star} \to \ell\ell\nu\nu$ decay channel has expected cross section of about 250 fb at 8 TeV for m_H =125 GeV. The search mass range is 110–200 GeV. The signature is two high-p_T isolated opposite-sign leptons and large E_T^{miss} due to neutrinos so one cannot reconstruct Higgs boson mass. Instead, its transverse mass reconstruction is possible. It is calculated as $m_T = \sqrt{(E_T^{\ell\ell} + E_T^{miss})^2 - |\mathbf{P_T^{\ell\ell}} + \mathbf{E_T^{miss}}|^2}$, where $E_T^{\ell\ell} = \sqrt{|P_T^{\ell\ell}|^2 + m_{\ell\ell}^2}$, $P_T^{\ell\ell}$ ($m_{\ell\ell}$) is transverse momentum (invariant-mass) of the di-lepton system, respectively. So knowledge of background normalization and shapes is of importance. Backgrounds come from $q\bar{q} + gg \to WW$ continuum production as well as from $t\bar{t}$, Wt and Z+jets processes with di-lepton final state. W+jets with a jet misidentified as a lepton has also sizeable rate. Separate control regions for each of these backgrounds are defined to measure them. Event sample is subdivided in accordance with number of jets (0, 1 or ≥2) and di-lepton flavour (ee, $\mu\mu$ or $e\mu$). For the 0 and 1 jet cases the

Table 1. For an assumed 0^+ hypothesis, the values for the expected and observed p_0-values of the different tested spin and parity hypotheses for the BDT and J^P-MELA analyses of the $H \to 4\ell$ channel. The results are given combining the 8 TeV and 7 TeV data sets. Also given is the observed p_0-value where 0^+ is the test hypothesis and the other spins states are the assumed hypothesis ("observed"). These two observed p_0-values are combined to provide the CL_s confidence level for each test hypothesis. The production mode is assumed to be 100% ggF.

	BDT analysis				MELA analysis			
	tested J^P for assumed 0^+		tested 0^+ for assumed J^P	CL_s	tested J^P for assumed 0^+		tested 0^+ for assumed J^P	CL_s
	expected	observed	observed		expected	observed	observed	
$0^- p_0$	0.0037	0.015	0.31	0.022	0.0011	0.0022	0.40	0.004
$1^+ p_0$	0.0016	0.001	0.55	0.002	0.0031	0.0028	0.51	0.006
$1^- p_0$	0.0038	0.051	0.15	0.060	0.0010	0.027	0.11	0.031
$2^+ p_0$	0.092	0.079	0.53	0.168	0.064	0.11	0.38	0.182
$2^- p_0$	0.0053	0.25	0.034	0.258	0.0032	0.11	0.08	0.116

ggF mechanism is expected to dominate while for the multi-jet case signal events are mostly due to the VBF mechanism. The m_T-distribution after background subtraction is shown in Fig.5(a). Clear excess of events is seen in the region 80–140 GeV. The local p_0 as function of m_H is shown in Fig.5(b). It corresponds to 3.8σ signal significance at 126 GeV. The fitted signal strength is equal to 1.0±0.3, i.e. compatible with the SM prediction. 95% CL of signal exclusion is reached (expected) in the region above 134 (118) GeV.

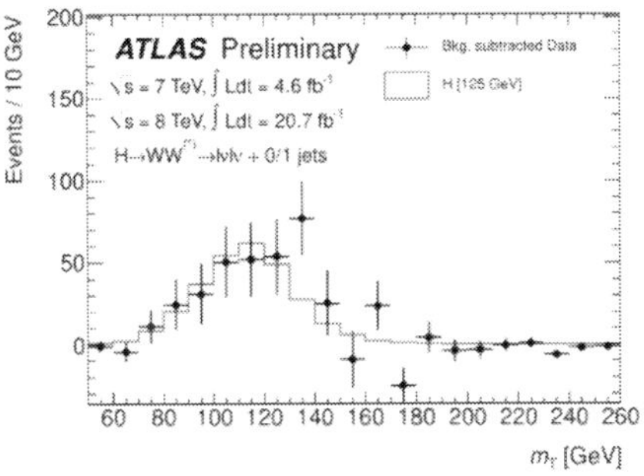

(a)

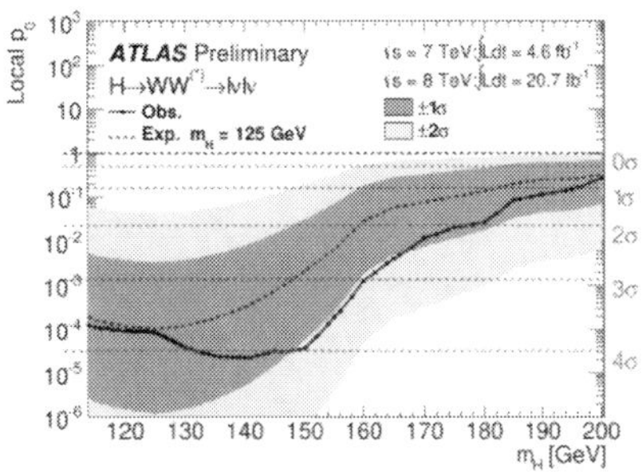

(b)

Figure 5. Results of the $H \to WW^\star \to \ell\ell\nu\nu$ study. (a) Background-subtracted m_T distribution. The signal is overlaid. The error bars represent the statistical uncertainties of the data and the subtracted background; it does not include the systematic uncertainties of the latter. (b) Results for the p_0 as function of m_H. The smaller green bands represent 1σ uncertainties on the expected values, and the larger yellow bands represent 2σ uncertainties.

Spin-parity properties are also studied with the $H \to WW^\star \to \ell\ell\nu\nu$ decay channel [11]. Two hypotheses were considered: 0^+ and 2^+. To discriminate between them, multivariate analysis based on four variables ($m_{\ell\ell}$, $\Delta\phi\ell\ell$, $P_T^{\ell\ell}$ and m_T), was performed.[5] Depending on the $f_{q\bar{q}}$ fraction,

[5] $\Delta\phi_{\ell\ell}$ is azimuthal angle gap between leptons.

exclusion of the 2^+ hypothesis at the CL between 95% and 99% is reached.

In addition, the search for $H \to WW \to \ell\ell\nu\nu$ in the high-mass region 260–1000 GeV is made [12]. The predicted $\sigma_H \times BR(H \to WW \to \ell\ell\nu\nu)$ at 8 TeV decreases from 300 fb at m_H =260 GeV to 4 fb at m_H =1000 GeV. Selection criteria were specially optimized to high-mass searches and a cut-based analysis is performed. Only $H \to WW \to e\nu\mu\nu$ channel is considered. Two extreme Higgs boson lineshapes are taken: the SM-like, i.e. from the Complex Pole Scheme (CPS), and zero width, i.e. from the Narrow Width Approximation (NWA). Corresponding exclusion plots are shown in Fig.6(a) and in Fig.6(b), respectively. SM-like Higgs boson with $m_H \leq 642$ GeV is excluded at 95% CL; a bit higher value is obtained for the NWA case.

At last, associated VH-production with subsequent $H \to WW \to \ell\ell\nu\nu$ decay is investigated in the range 110 GeV$\leq m_H \leq$200 GeV [13]. To have better S/B ratio, Z (W) bosons are tagged via their $\ell\ell$ ($\ell\nu$) decays, respectively. Expected signal cross section is about 1.8 fb at m_H =125 GeV. The signature is three or four isolated, high-p_T leptons and large E_T^{miss}. Backgrounds include WWW-, and WZ/γ^*- and ZZ-production as well as other di-lepton final states with additional fake lepton. The study has shown that $\sigma_H \leq 7.2\sigma_{SM}$ is excluded at 95% CL for m_H =125 GeV. With the inclusion of VH-production, the combined $H \to WW \to \ell\ell\nu\nu$ signal significance increases from 3.8σ to 4.0σ.

3.4. VH, $H \to b\bar{b}$ decay channel [14]

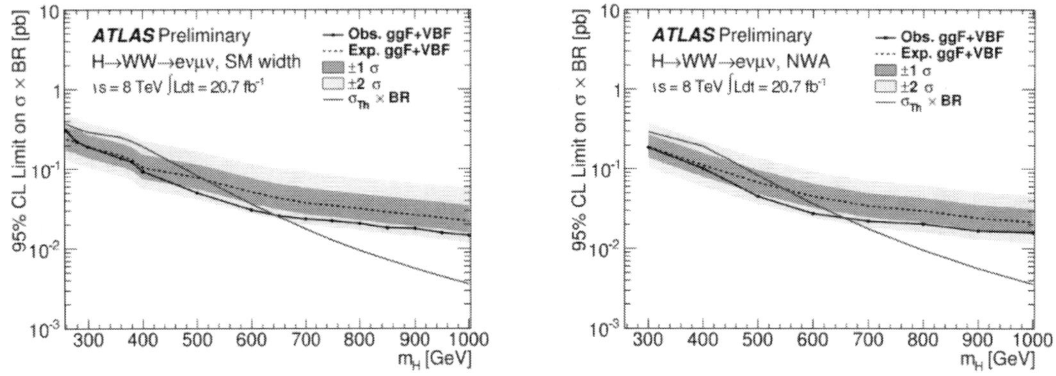

(a) (b)

Figure 6. 95% CL upper limits on $\sigma_H \times BR(H \to WW \to e\nu\mu\nu)$ at 8 TeV for a Higgs boson with (a) a SM-like lineshape and (b) a narrow lineshape (NWA). The green and yellow bands show the 1σ and 2σ uncertainties on the expected limit. The expected cross section times branching ratio for the production of a SM Higgs boson is shown as a blue line.

Among all channels, $H \to b\bar{b}$ is predicted to have the largest rate in the mass range 110–150 GeV. However, background is known to be huge. For this reason, here associated VH-production is of interest. The signature includes two energetic jets originating from b-quarks having invariant-mass $m_{b\bar{b}}$ close to m_H as well as one or two high-p_T leptons or large E_T^{miss} depending on the decay mode of the associated vector boson ($W \to \ell\nu$, $Z \to \ell\ell$, $Z \to \nu\nu$, respectively). Backgrounds consist of $W/Z+$ jets, dibosons, top and multijet production with fake leptons. Reconstructed $m_{b\bar{b}}$ is shown in Fig.7(a) after subtraction of all backgrounds except dibosons. No significant deviation of the data from simulations is seen. $1.4\sigma_{SM}$ cross section is excluded at 95% CL for m_H =125 GeV, see Fig.7(b). The measured signal strength is μ =0.2±0.5(stat)±0.4(sys).

4. Statistical combination of $H \to \gamma\gamma$, $H \to ZZ^\star \to 4\ell$, $H \to WW^\star \to \ell\ell\nu\nu$, $H \to b\bar{b}$ and $H \to \tau\tau$ decay channels

4.1. Signal strength, reconstructed mass and measured couplings [15], [16], [17]

The local probability p_0 for a background-only experiment to be more signal-like than the observation as function of m_H for the combination of all channels is shown in Fig.8(a). The observed local significance is 10σ and the expected significance at m_H =125 GeV is 7.5σ [16]. The measured signal mass based on $H \to \gamma\gamma$ and $H \to ZZ^\star \to 4\ell$ channels is m_H^{obs} =125.5±0.2±0.6 GeV with 2.5σ tension between the values predicted from these separate channels [15][6]. The combined signal strength in units of the σ_{SM}, μ, is measured to be 1.33±0.14(stat)±0.15(sys) from three channels $H \to \gamma\gamma$, $H \to ZZ^\star \to 4\ell$ and $H \to WW^\star \to \ell\ell\nu\nu$. If the preliminary $H \to \tau\tau$ [18] and $H \to b\bar{b}$ [19] results, for which only part of the 8 TeV data is used (13 fb^{-1}), were included, the combined signal strength would be μ =1.23±0.18.

Since several Higgs boson production modes are available at the LHC, one can measure separately two signal strength parameters in units of corresponding SM-predicted cross sections, $\mu_{ggF+ttH} = \mu_{ggF} = \mu_{ttH}$ from gluon-mediated processes and $\mu_{VBF+VH} = \mu_{VBF} = \mu_{VH}$ from vector-boson-mediated processes. A ratio $\mu_{VBF+VH}/\mu_{ggF+ttH}$ $=1.4^{+0.7}_{-0.5}$ is obtained from the combination of three channels, $H \to \gamma\gamma$, $H \to WW^\star \to \ell\ell\nu\nu$ and $H \to ZZ^\star \to 4\ell$. This result provides evidence at the 3.3σ level that a fraction of H production occurs through the VBF.

Measurements of couplings are implemented using a leading-order tree-level motivated framework with some assumptions [17]. The coupling scale factors κ_j are defined in such a way that the cross sections σ_j and the partial decay widths Γ_j associated with the SM particle j scale with κ^2 compared to the SM prediction. Assuming one coupling scale factor for fermions, κ_F, and one for bosons, κ_V, the 68% CL intervals profiling over the other parameter, are $\kappa_F \in$[0.76,1.18] and $\kappa_V \in$[1.05,1.22]. If we relax the assumption that there are no contributions from new particles to the Higgs boson width, only the ratio $\lambda_{FV} = \kappa_F/\kappa_V$ can be measured which still provides useful information on the relationship between Yukawa and gauge couplings. Fits to the data give the following 68% CL intervals for λ_{FV} and $\kappa_{VV} = \kappa_V\kappa_V/\kappa_H$: $\lambda_{FV} \in$[0.70,1.01] and $\kappa_{VV} \in$[1.13,1.45].

Many BSM physics scenarios predict the existence of new heavy particles, which can contribute to loop-induced processes such as ggF production and $H \to \gamma\gamma$ decay. Effective scale factors κ_g and κ_γ are introduced to parameterise the ggF and $H \to \gamma\gamma$ loops. The best-fit values when profiling over the other parameters are: κ_g =1.04±0.14 and κ_γ =1.20±0.15.

In summary, no significant deviation from the SM prediction is observed in any of the measurements performed.

4.2. Spin combination [20], [21]

The three H decay channels are combined to study spin properties of the observed new boson: $H \to \gamma\gamma$, $H \to ZZ^\star \to 4\ell$ and $H \to WW^\star \to e\nu\mu\nu$. Complete 8-TeV dataset is used for this study; for the $H \to ZZ^\star \to 4\ell$ channel all data obtained at 7 TeV are added. Five different spin-parity hypotheses were tested: 0^+, 0^-, 1^+, 1^-, and 2^+. Data strongly favour the 0^+ hypothesis. 1^+ and 2^+ hypotheses are excluded with CL≥99.9%, see Fig.8(b). Table 2 shows the expected and observed p_0 values for both the 0^+ and 2^+ hypotheses for the combination of mentioned channels. The results are shown as function of the $f_{q\bar{q}}$ production fraction.

[6] $\Delta m_H = m_H(\gamma\gamma) - m_H(4\ell)$ =2.3±0.7(stat)±0.6(sys) GeV.

Table 2. Expected and observed p_0 values for the 0^+ and 2^+ hypotheses as function of the $f_{q\bar{q}}$ of the spin-2 production mechanism in the combined H study. The values are calculated for the combination of the three H decay channels. The CLs values are also presented.

$f_{q\bar{q}}$	2^+ assumed exp. $p_0(J^P=0^+)$	0^+ assumed exp. $p_0(J^P=2^+)$	obs. p_0 $(J^P=0^+)$	obs. p_0 $(J^P=2^+)$	CL_s $(J^P=2^+)$
100%	3.0×10^{-3}	8.8×10^{-5}	0.81	1.6×10^{-6}	0.8×10^{-5}
75%	9.5×10^{-3}	8.8×10^{-4}	0.81	3.2×10^{-5}	1.7×10^{-4}
50%	1.3×10^{-2}	2.7×10^{-3}	0.84	8.6×10^{-5}	5.3×10^{-4}
25%	6.4×10^{-3}	2.1×10^{-3}	0.80	0.9×10^{-4}	4.6×10^{-4}
0%	2.1×10^{-3}	5.5×10^{-4}	0.63	1.5×10^{-4}	4.2×10^{-4}

5. Conclusion

- Based on analysis of $H\to\gamma\gamma$, $H\to ZZ^\star\to 4\ell$, $H\to WW^\star\to\ell\ell\nu\nu$ decay channels in complete dataset taken at $\sqrt{s}=7$ TeV and 8 TeV, the ATLAS experiment discovered neutral boson with a mass $\approx$125 GeV which has production cross section compatible with the SM Higgs boson; fermionic decay modes of this boson are not yet discovered;
- First measurements of the new boson couplings were performed and they are all in agreement with the SM predictions; coupling to bosons were measured with much better precision than couplings to fermions;
- Different spin/parity hypotheses of the new particle were tested and the one predicted by the SM, $J^P=0^+$ hypothesis, has strong preference;
- ATLAS is continuing to study properties of the discovered boson and plan to improve the measurements with new data expected starting from year 2015. An overview of the expected measurement precision in each channel for the signal strength μ and its coupling constants at 14-TeV LHC with respect to the SM H with $m_H=125$ GeV is given in Ref.[22].

[1] Dittmaier S et al, http://arxiv.org/abs/1101.1593
[2] ATLAS Collaboration, 2012 *Phys. Lett. B* **716** 1.
[3] CMS Collaboration, 2012 *Phys. Lett. B* **716** 30.
[4] ATLAS Collaboration, 2008 JINST **3** S08003.
[5] Allison J et al, 2006 *IEEE Trans. Nucl. Sci.* **53** 270; Agostinelli S et al, 2003 *NIM A* **506** 250.
[6] ATLAS Collaboration, 2013 ATLAS-CONF-2013-012, http://cds.cern.ch/record/1523698
[7] ATLAS Collaboration, 2013 ATLAS-CONF-2013-029, http://cds.cern.ch/record/1527124
[8] ATLAS Collaboration, 2012 ATLAS-CONF-2012-168, http://cds.cern.ch/record/1499625
[9] ATLAS Collaboration, 2013 ATLAS-CONF-2013-013, http://cds.cern.ch/record/1523699
[10] ATLAS Collaboration, 2013 ATLAS-CONF-2013-030, http://cds.cern.ch/record/1527126
[11] ATLAS Collaboration, 2013 ATLAS-CONF-2013-031, http://cds.cern.ch/record/1527127
[12] ATLAS Collaboration, 2013 ATLAS-CONF-2013-067, http://cds.cern.ch/record/1562879
[13] ATLAS Collaboration, 2013 ATLAS-CONF-2013-075, http://cds.cern.ch/record/1562933
[14] ATLAS Collaboration, 2013 ATLAS-CONF-2013-079, http://cds.cern.ch/record/1563235
[15] ATLAS Collaboration, 2013 ATLAS-CONF-2013-014, http://cds.cern.ch/record/1523727
[16] ATLAS Collaboration, 2013 ATLAS-CONF-2013-034, http://cds.cern.ch/record/1528170
[17] ATLAS Collaboration, 2013 *Phys. Lett. B* **726** 88. http://arxiv.org/pdf/1307.1427
[18] ATLAS Collaboration, 2012 ATLAS-CONF-2012-160, http://cds.cern.ch/record/1493624
[19] ATLAS Collaboration, 2012 ATLAS-CONF-2012-161, http://cds.cern.ch/record/1493625
[20] ATLAS Collaboration, 2013 ATLAS-CONF-2013-040, http://cds.cern.ch/record/1542341
[21] ATLAS Collaboration, 2013 *Phys. Lett. B* **726** 120, http://arxiv.org/pdf/1307.1432
[22] ATLAS Collaboration, 2012 ATLAS-PHYS-PUB-2012-004.

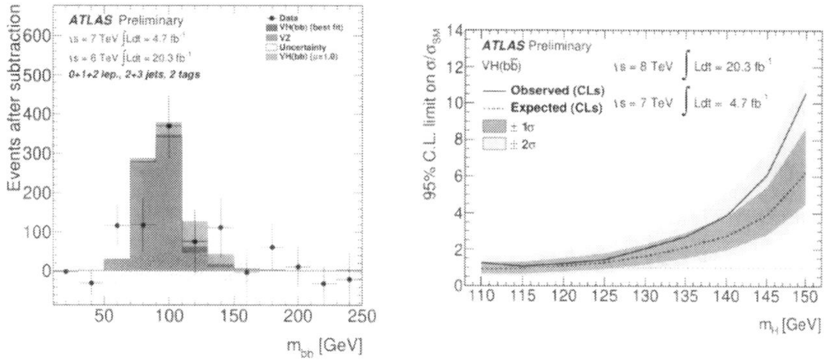

(a) (b)

Figure 7. Results of study of associated VH-production with subsequent $H \to b\bar{b}$ decay. (a) The $m_{b\bar{b}}$ distribution in data after subtraction of all backgrounds except for the diboson processes. The Higgs boson signal contribution is shown both with its fitted signal strength (indicated as "best fit") and as expected for the SM cross section (indicated as μ =1.0). The size of the combined statistical and systematic uncertainty on the fitted background is indicated by the hashed band. (b) Expected (dashed) and observed (solid) 95% CL cross section upper limits, normalised to the σ_{SM}, as function of m_H for all channels and data taking periods combined. The expected upper limit is given for the background-only hypothesis. The green and yellow bands represent the 1σ and 2σ ranges of the expectation in the absence of a signal.

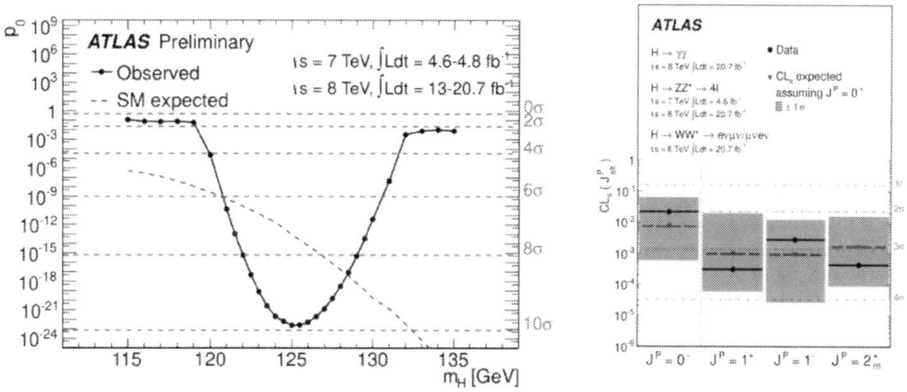

(a) (b)

Figure 8. (a) The local probability p_0 for a background-only experiment to be more signal-like than the observation as function of m_H for the combination of all H channels. The dashed curve shows the median expected local p_0 under the hypothesis of a SM Higgs boson production at that mass. (b) Expected (blue triangles/dashed lines) and observed (black circles/solid lines) CL_s for alternative spin-parity hypotheses assuming a 0^+ signal. The green band represents the 68% CL_s expected exclusion range for a signal with assumed 0^+. For the spin-2 hypothesis, the results for the specific 2^+ model, are shown. On the right y-axis, the corresponding numbers of Gaussian standard deviations are given, using the one-sided convention.

SUSY dark matter annihilation in the Galactic halo

Veniamin Berezinsky, Vyachelav Dokuchaev and Yury Eroshenko
Institute for Nuclear Research, Russian Academy of Sciences,
prospekt 60-letiya Oktyabrya 7a, 117312 Moscow, Russia

E-mail: berezinsky@lngs.infn.it, dokuchaev@inr.ac.ru, eroshenko@inr.ac.ru

Abstract. Neutralino annihilation in the Galactic halo is the most definite observational signature proposed for indirect registration of the SUSY Dark Matter (DM) candidate particles. The corresponding annihilation signal (in the form of gamma-rays, positrons and antiprotons) may be boosted for one or three orders of magnitude due to the clustering of cold DM particles into the small-scale and very dense self-gravitating clumps. We discuss the formation of these clumps from the initial density perturbations and their successive fate in the Galactic halo. Only a small fraction of these clumps, ~ 0.1 %, in each logarithmic mass interval $\Delta \log M \sim 1$ survives the stage of hierarchical clustering. We calculate the probability of surviving the remnants of dark matter clumps in the Galaxy by modelling the tidal destruction of the small-scale clumps by the Galactic disk and stars. It is demonstrated that a substantial fraction of clump remnants may survive through the tidal destruction during the lifetime of the Galaxy. The resulting mass spectrum of survived clumps is extended down to the mass of the core of the cosmologically produced clumps with a minimal mass. The survived dense remnants of tidally destructed clumps provide an amplification (boosting) of the annihilation signal with respect to the diffuse DM in the Galactic halo. We describe the anisotropy of clump distribution caused by the tidal destruction of clumps in the Galactic disk.

1. Introduction

The stable SUSY particle like neutralino is a promising candidate for the enigmatic Dark Matter (DM) particle. A primordial power-law spectrum of density fluctuations in the DM ranges from the largest scales above the scales of superclusters of galaxies to the smallest sub-stellar scales according to prediction of inflation models. This permits to predict the properties of smallest DM structures from the known Cosmic Microwave Background (CMB) fluctuations at large scales. The most bright indirect signature of the SUSY DM particles is their annihilation, and this process could be boosted inside the dense DM clumps in the galactic halo.

The cosmological formation and evolution of small-scale DM clumps have been studied in numerous works [1]–[16]. The minimum mass of clumps (the cutoff of the mass spectrum), $M_{\rm min}$ is determined by the collision and collisionless damping processes (see, e. g., [5] and references therein). Additionally the cutoff of mass spectrum is influenced by the acoustic absorption [17] at the time of kinetic decoupling of the SUSY DM particles [18] and also by the horizon-scale perturbation modes [19]. The low-mass cut-off of the clump mass-spectrum accompanies the process of decoupling. It starts when DM particles coupled strongly with surrounding plasma in the growing density fluctuations. The smearing of the small-scale fluctuations is due to the collision damping occurring just before decoupling, in analogy with the Silk damping [20]. It occurs due to diffusion of DM particles from a growing fluctuation, and only the small-scale fluctuations can be destroyed by this process. The corresponding diffusive cut-off $M_{\rm min}^{\rm diff}$ is very small. As coupling becomes weaker, the larger fluctuations are destroyed and $M_{\rm min}$ increases. One may expect that the largest value of $M_{\rm min}$ is related to a free-streaming regime. However, as recent calculations show [9], the largest $M_{\rm min}$ is related to some friction between DM particles and cosmic plasma similar to the Silk damping. The predicted minimal clump masses range from very low values, $M_{\rm min} \sim 10^{-12} M_{\odot}$ [21], produced by diffusive escape of DM particles, up to $M_{\rm min} \sim 10^{-4} M_{\odot}$, caused by acoustics oscillations [19] and quasi-free-streaming with limited friction [9]. In the case of the Harrison-Zeldovich spectrum of primordial fluctuations with CMB normalization the first Earth-mass small-scale DM clumps are formed at redshift $z \sim 60$ (for 2σ fluctuations) with a mean density 7×10^{-22} g cm^{-3}, virial radius 6×10^{-3} pc and internal velocity dispersion 80 cm s^{-1} respectively. Only very small fraction of these clumps survives the early stage of tidal destruction during the hierarchial clustering [3]. Nevertheless these survived clumps may provide the major contribution to the annihilation signal in the Galaxy [3, 6, 22, 23, 24]. At a high redshift the SUSY neutralinos, considered as DM particles, may cause the efficient heating of the diffuse gas [25] due to annihilation in the dense clumps.

One of the unresolved problem of DM clumps is a value of the central density or core radius. Numerical simulations give a nearly power density profile of DM clumps. Both the Navarro-Frenk-White (NFW) and Moore profiles give formally a divergent density in the clump center. A theoretical modelling of the clump formation [26, 27, 28] predicts a power-law profile of the internal density of clumps

$$\rho_{\rm int}(r) = \frac{3-\beta}{3}\bar{\rho}\left(\frac{r}{R}\right)^{-\beta}, \tag{1}$$

where $\bar{\rho}$ and R are the mean internal density and a radius of clump, respectively, $\beta \simeq 1.8 - 2$ and $\rho_{\rm int}(r) = 0$ at $r > R$. A near isothermal power-law profile (1) with $\beta \simeq 2$ has been recently obtained in numerical simulations of small-scale clump formation [29]. In [3] the core radius $x_c \simeq 0.3\nu^{-2}$ has been obtained, where ν is a relative height of the fluctuation density peak in units of dispersion at the time of energy-matter equality. This value is a result of the influence of tidal forces on the motion of DM particles in the clump at stage of formation. This estimate may be considered as an upper limit for the core radius or as the break-scale in the density profile, e. g., a characteristic scale in the Navarro-Frenk-White profile. It could be that a real core radius, where the density ceases to grow, is determined by the relaxation of small-scale perturbations inside the forming clump [30]. Another mechanism for core formation arises in the 'meta-cold dark matter model' due to late decay of cold thermal relics into lighter nonrelativistic particles with low phase-space density [31, 32]. Here we consider the relative core radius $x_c = R_c/R$ of DM clumps as a free parameter in the range $0.001 - 0.1$. Correct approach includes a gradual mass loss of a systems [33, 34, 36], in particular, by small-scale DM clumps [4, 35].

2. Formation and Destruction of clumps in hierarchial clustering

The process of hierarchical clustering and tidal destruction of DM clumps can be outlined in the following way. The DM clumps of minimal mass are formed first in the expanding Universe. The clumps of larger mass, which host the smaller ones are formed later, and so on. Some part of clumps are destroyed tidally in the gravitational field of their host clumps. In this Section we study the destruction of DM clumps in the process of hierarchical structuring long before the final galaxy formation. At small-mass scales the hierarchial clustering is a fast and rather complicated nonlinear process. A formation time of clump with an internal density ρ is $t = (\kappa\rho_{\rm eq}/\rho)^{1/2} t_{\rm eq}$, where $\kappa = 18\pi^2$ and $\rho_{\rm eq} = \rho_0(1+z_{\rm eq})^3$ is a cosmological density at the time of matter-radiation equality $t_{\rm eq}$, $1 + z_{\rm eq} = 2.35 \times 10^4 \Omega_m h^2$ and $\rho_0 = 1.9 \times 10^{-29} \Omega_m h^2$ g cm^{-3}. The index 'eq' here and throughout below refers to quantities at the time of matter-radiation equality $t_{\rm eq}$. The DM clumps of mass M can be formed from density fluctuations of different peak-height $\nu = \delta_{\rm eq}/\sigma_{\rm eq}(M)$, where $\sigma_{\rm eq}(M)$ is the fluctuation dispersion on a mass-scale M at the time $t_{\rm eq}$. A mean internal density of clump ρ is fixed at the time of clump formation and according to [38] equals $\rho = \kappa\rho_{\rm eq}[\nu\sigma_{\rm eq}(M)/\delta_c]^3$, where $\delta_c = 3(12\pi)^{2/3}/20 \simeq 1.686$.

An internal energy of self-gravitating object increases in tidal interactions. This energy increase was calculated, e. g., in [39] for the case of a star globular cluster in a spherical galaxy. By using the model of tidal heating from [39], we determine now a survival time (or a time of tidal destruction) T of some chosen small-scale clump due to the tidal heating inside of a larger mass host clump. The motion of a clump would be rather complicated in the case of a fast hierarchical clustering of hosts. During a dynamical time in the host $t_{\rm dyn} \simeq 0.5(G\rho_h)^{-1/2}$, where ρ_h is a mean internal density of the host, the chosen small-scale clump can belong to several successively destructed hosts. A clump trajectory in the host experiences successive turns accompanied by the "tidal shocks" [40, 39]. For the considered small-scale clump with a mass M and radius R, the corresponding internal energy increase after a single tidal shock is

$$\Delta E \simeq \frac{4\pi}{3}\gamma_1 G\rho_h M R^2, \tag{2}$$

where a numerical factor $\gamma_1 \sim 1$. Let us denote the number of tidal shocks per dynamical time $t_{\rm dyn}$ by γ_2. A corresponding rate of clump internal energy growth is $\dot{E} = \gamma_2\Delta E/t_{\rm dyn}$. A clump is destroyed in the host if its internal energy increase due to tidal shocks exceeds a total energy $|E| \simeq GM^2/2R$. As a result, for a typical time $T = T(\rho, \rho_h)$ of the tidal destruction of a small-scale clump with density ρ inside a more massive host with a density ρ_h we obtain:

$$T^{-1}(\rho, \rho_h) = \dot{E}/|E| \simeq 4\gamma_1\gamma_2 G^{1/2}\rho_h^{3/2}\rho^{-1}. \tag{3}$$

It turns out that a resulting mass function of small-scale clumps (see in this Section below) depends rather weakly on the value of $\gamma_1\gamma_2$.

The probability of clump survival, determined as a fraction of the clumps with mass M surviving the tidal destruction in hierarchical clustering, is given by the exponential function e^{-J} with

$$J \simeq \sum_h \frac{\Delta t_h}{T(\rho, \rho_h)}. \tag{4}$$

Here Δt_h is a difference of formation times t_h for two successive hosts, and summation goes over all clumps of intermediate mass-scales, which successively host the considered small-scale clump of a mass M. Changing the summation by integration in (4) we obtain

$$J(\rho, \rho_f) = \int\limits_{t_1}^{t_f} \frac{dt_h}{T(\rho, \rho_h)} \simeq \gamma\frac{\rho_1 - \rho_f}{\rho} \simeq \gamma\frac{\rho_1}{\rho} \simeq \gamma\frac{t^2}{t_1^2}, \tag{5}$$

where

$$\gamma = 2\gamma_1\gamma_2\kappa^{1/2}G^{1/2}\rho_{\rm eq}^{1/2}t_{\rm eq} \simeq 14(\gamma_1\gamma_2/3), \tag{6}$$

and t, t_1, t_f, ρ, ρ_1 and ρ_f are respectively the formation times and internal densities of the considered clump and of its first and final hosts. One may see from Eq. (5) that the first host provides a major contribution to the tidal destruction of the considered small-scale clump, especially if the first host density ρ_1 is close to ρ, and consequently $e^{-J} \ll 1$.

Now we need to track the number of clumps M (originated from the density peak ν) which enter some larger host during time intervals Δt_1 around each t_1 beginning from the time t of clump formation. A mass function of small-scale clumps (i. e., a differential mass fraction of DM in the form of clumps survived in hierarchical clustering) can be expressed as

$$\xi\frac{dM}{M}d\nu = dM\,d\nu\,\frac{e^{-\nu^2/2}}{\sqrt{2\pi}}\int\limits_{t(\nu\sigma_{\rm eq})}^{t_0} dt_1 \left|\frac{\partial^2 F(M, t_1)}{\partial M\,\partial t_1}\right| e^{-J(t,t_1)}. \tag{7}$$

In this expression t_0 is the Universe age and $F(M,t)$ is a mass fraction of unconfined clumps (i. e., clumps, not belonging to more massive hosts) with a mass smaller than M at time t. According to [38], the mass fraction of unconfined clumps is $F(M,t) = \text{erf}\left(\delta_c/[\sqrt{2}\sigma_{\text{eq}}(M)D(t)]\right)$, where $\text{erf}(x)$ is the error-function and $D(t)$ is the growth factor normalized by $D(t_{\text{eq}}) = 1$. An upper limit of integration t_0 in Eq. (7) is not crucial and may be extrapolated to infinity because a main contribution to the tidal destruction of clumps is provided by the early formed hosts at the beginning of the hierarchical clustering. Two processes are responsible for time evolution of the fraction $\partial^2 F/(\partial M \partial t)$ for unconfined clumps in the mass interval dM: (i) the formation of new clumps and (ii) the capture of smaller clumps into the larger ones. Both these processes are equally efficient at the time when $\partial^2 F/(\partial M \partial t) = 0$. Finally, we transform the distribution function (7) to the following form:

$$\xi \frac{dM}{M} d\nu \simeq \frac{\nu\, d\nu}{\sqrt{2\pi}} e^{-\nu^2/2} f_1(\gamma) \frac{d \log \sigma_{\text{eq}}(M)}{dM} dM, \tag{8}$$

where

$$f_1(\gamma) = \frac{2[\Gamma(1/3) - \Gamma(1/3,\gamma)]}{3\sqrt{2\pi}\gamma^{1/3}}. \tag{9}$$

Here $\Gamma(1/3)$ and $\Gamma(1/3,\gamma)$ are the Euler gamma-function and incomplete gamma-function, respectively. The function $f_1(\gamma)$ varies rather slowly in the interesting interval of $14 < \gamma < 40$, and one may use $f_1(\gamma) \simeq 0.2 - 0.3$. Physically the first factor ν in (8) corresponds to a more effective survival of high-density clumps (i. e., with large values of ν) with respect to the low-density ones (with small values of ν). Integrating Eq. (8) over ν, we obtain

$$\xi_{\text{int}} \frac{dM}{M} \simeq 0.02(n+3) \frac{dM}{M}. \tag{10}$$

An effective power-law index n in Eq. (10) is determined as $n = -3(1 + 2\partial \log \sigma_{\text{eq}}(M)/\partial \log M)$ and depends very weakly on M.

In further calculations we use an interpolation fitting of the fluctuation dispersion $\sigma_{\text{eq}}(M)$ from [41] and [1]. The analysis of the WMAP data of the CMB anisotropy [42] reveals a power-law spectrum of initial perturbations with $n_p = 0.99 \pm 0.04$ in a good agreement with the canonical inflation value $n_p = 1.0$. However, when the data from 2dF galaxy power-spectrum and Ly-α are included in the analysis, the best-fit favors in a softer spectrum with $n_p = 0.96 \pm 0.02$.

The simple M^{-1} shape of the mass function (10) is in a very good agreement with the corresponding numerical simulations [29], but our normalization factor is a few times smaller. It must be stressed that a physical meaning of the survived clump distribution function $\xi\, d\nu\, dM/M$ is different from the similar one for the unconfined clumps, given by the Press-Schechter mass function $\partial F/\partial M$.

3. Tidal destruction of clumps by the Galactic disk and stars

Crossing the Galactic disc, a clump can be tidally destructed by the collective gravitational field of stars in the disc. This phenomenon is similar to the destruction of a globular cluster by the "tidal shocking" in the Galactic disc [43]. The kinetic energy gain of a DM particle with respect to the center of clump after one crossing of the Galactic disk is [43]

$$\delta E = \frac{4g_m^2(\Delta z)^2 m}{v_{z,c}^2} A(a), \tag{11}$$

where m is a constituent DM particle mass, Δz is a vertical distance (orthogonal to the disk plane) of a DM particle with respect to the center of clump, $v_{z,c}$ is a vertical velocity of clump with respect to the disk plane at the moment of disk crossing and $A(a)$ is the adiabatic correction factor. A gravitational acceleration near the disk plane is $g_m(r) = 2\pi G\sigma_s(r)$, where we use an exponential model for a surface density of disk. The factor $A(a)$ in (11) describes the adiabatic protection from slow tidal effects [44]. In [33] the following fitting formula was proposed: $A(a) = (1 + a^2)^{-3/2}$. Here the adiabatic parameter $a = \omega\tau_d$, where ω is an orbital frequency of DM particle in the clump, $\tau_d \simeq H_d/v_{z,c}$ is an effective duration of gravitational tidal shock produced by the disk with a half-thickness H_d.

As a representative example we consider the isothermal internal density profile of DM clump

$$\rho_{\rm int}(r) = \frac{1}{4\pi}\frac{v_{\rm rot}^2}{Gr^2} \tag{12}$$

with a cutoff at the virial radius R: $\rho(r) = 0$ at $r > R$. A corresponding mass profile of clump is $M(r) = M_i(r/R)$, where M_i is an initial mass of clump at the epoch of Galaxy formation. With this mass distribution a circular velocity inside a clump is independent of radius, $v_{\rm rot} = (GM(r)/r)^{1/2} = (GM_i/R)^{1/2}$. A gravitational potential corresponding to the density profile (12) is $\phi(r) = v_{\rm rot}^2[\log(r/R) - 1]$. Let us define a dimensionless energy of the DM particle $\varepsilon = E/(mv_{\rm rot}^2)$ and gravitational potential $\psi(r) = \phi(r)/v_{\rm rot}^2 = \ln(r/R) - 1$. An internal density profile $\rho_{\rm int}(r)$ and the distribution function of DM particles in the clump $f_{\rm cl}(\varepsilon)$ are related by the integral relation [45]

$$\rho_{\rm int}(r) = 2^{5/2}\pi \int\limits_{\psi(r)}^{0} \sqrt{\varepsilon - \psi(r)}\, f_{\rm cl}(\varepsilon)\, d\varepsilon, \tag{13}$$

The corresponding isothermal distribution function is $f_{\rm cl}(\varepsilon) \propto \exp(-2\varepsilon)$.

By using the hypothesis of a tidal stripping of outer layers of a DM clump, we see that a tidal energy gain $\delta\varepsilon$ causes the stripping of particles with energies in the range $-\delta\varepsilon < \varepsilon < 0$. A corresponding variation of density at radius r is

$$\delta\rho(r) = 2^{5/2}\pi \int\limits_{-\delta\varepsilon}^{0} \sqrt{\varepsilon - \psi(r)}\, f_{\rm cl}(\varepsilon)\, d\varepsilon. \tag{14}$$

In this equation the tidal energy gain (11) by different DM particles is averaged over angles, so as $\langle(\Delta z)^2\rangle = r^2/3$. A resulting total mass loss by DM clump during one crossing of the Galactic disk is

$$\delta M = -4\pi \int_0^R r^2 \delta\rho(r)\, dr. \tag{15}$$

Let us specify the dimensionless quantities

$$Q_d = \frac{g_m^2}{2\pi v_{z,c}^2 G\bar{\rho}_i}, \quad S_d = \frac{4\pi}{3} G\bar{\rho}_i \tau_d^2, \tag{16}$$

where $\bar{\rho}_i = 3M_i/(4\pi R^3)$ is a initial mean density of clump. For the most parts of clumps $Q_d \ll 1$ with a typical value $Q_d \sim 0.03$. The fitting formula for the mass loss of clump during one passage through the Galactic disk is

$$\left(\frac{\delta M}{M}\right)_d \simeq -0.13 Q_d \exp\left(-1.58 S_d^{1/2}\right). \tag{17}$$

Let us choose some particular clump moving in the spherical halo with an orbital “inclination” angle γ between the normal vectors of the disk plane and orbit plane. The orbit angular velocity at a distance r from the Galactic center is $d\phi/dt = J/(mr^2)$, where J is an orbital angular momentum of a clump. A vertical velocity of a clump crossing the disk is

$$v_{z,c} = \frac{J}{mr_s}\sin\gamma, \tag{18}$$

where r_s is a radial distance of crossing point from the Galaxy center. There are two crossing points (with different values of r_s) during an orbital period.

The standard Navarro-Frenk-White profile of the DM Galactic halo is

$$\rho_{\rm H}(r) = \frac{\rho_0}{(r/L)\,(1+r/L)^2}, \tag{19}$$

where $L = 45$ kpc, $\rho_0 = 5\times 10^6 {\rm M}_\odot$ kpc^{-3}. It useful to introduce the dimensionless variables:

$$x = \frac{r}{L}, \quad \tilde{\rho}_{\rm H}(x) = \frac{\rho_{\rm H}(r)}{\rho_0}, \quad y = \frac{J^2}{8\pi G\rho_0 L^4 M^2}, \quad \varepsilon = \frac{E_{\rm orb}/M - \Phi_0}{4\pi G\rho_0 L^2}, \quad \psi = \frac{\Phi - \Phi_0}{4\pi G\rho_0 L^2}, \tag{20}$$

where $\Phi_0 = -4\pi G\rho_0 L^2$, $E_{\rm orb}$ is a total orbital energy of a clump. With these variables the density profile of the halo (19) is written as $\tilde{\rho}_{\rm H}(x) = [x(1+x)^2]^{-1}$. A gravitational potential $\psi(x)$, corresponding to density profile (19) is $\psi(x) = 1 - \log(1+x)/x$. The relation between density profile $\tilde{\rho}_H(x)$ and the distribution function is given by the same equation (13) with an obvious substitution $f_{\rm cl} \Rightarrow F(\varepsilon)$, where the distribution function $F(\varepsilon)$ for a halo profile (19) can be fitted as in [46].

Let us denote the interval of time for motion from $x_{\rm min}$ to $x_{\rm max}$ and back as $T_c(x,\varepsilon,p)$, and the angle of orbital precession during the time $T_c/2$ as $\tilde{\phi}$. The orbital period is longer than T_c and is given by $T_t = T_c\left(1+\tilde{\phi}/\pi\right)^{-1}$. Choosing a time interval ΔT much longer than a clump orbital period T_t, but much shorter than the age of the Galaxy t_0, i. e., $T_t \ll \Delta T \ll t_0$, we define an averaged rate of mass loss by a selected clump under influence of tidal shocks in successive disk crossings:

$$\frac{1}{M}\left(\frac{dM}{dt}\right)_d \simeq \frac{1}{\Delta T}\sum\left(\frac{\delta M}{M}\right)_d, \tag{21}$$

where $(\delta M/M)_d$ is given by (17) and summation goes over all successive crossing points (odd and even) of a clump orbit with the Galactic disk during the time interval ΔT. The values of the g_m and $v_{z,c}$ both depend on the radius $x = r/L$. One simplification in calculation of (21) follows from the fact that a velocity of orbit precession is constant. For this reason the points of successive odd crossings are separated by the same angles $\tilde{\phi}$. The same is also true for successive even crossings. Using this simplification we transform the summation in (21) to integration:

$$\frac{1}{\Delta T}\sum\left(\frac{\delta M}{M}\right)_d \simeq \frac{2}{T_t|\tilde{\phi}|}\int\limits_{x_{\rm min}}^{x_{\rm max}}\left(\frac{\delta M}{M}\right)_d \frac{d\phi}{dx}dx,$$

where $d\phi/dx$ is an equation for clump orbit in the halo.

During a single close encounter of a DM clump with a star, the energy gain of a constituent DM particle in the clump with respect to clump center is [6]:

$$\delta E = \frac{2G^2 m_s^2 m\Delta z^2}{v_{\rm rel}^2 l^4}, \tag{22}$$

where m_* is a star mass, l is an impact parameter, $v_{\rm rel}$ is a relative star velocity with respect to a clump, $\Delta z = r\cos\psi$, r is a radial distance of a DM particle from the clump center and ψ

is an angle between the directions from the clump center to the DM particle and to the point of closest approach of a star. Using the same method as in the Section 3 we calculate a relative mass loss by clump $(\delta M/M)_s$ during a single encounter with a star and obtain the same fitting formula as (17) but with a substituting the dimensionless parameters, $Q_d \Rightarrow Q_s$ and $S_d \Rightarrow S_s$, where

$$Q_s = \frac{Gm_*^2}{2\pi v_{\rm rel}^2 l^4 \bar{\rho}_i}, \quad S_s = \frac{4\pi}{3} G\bar{\rho}_i \tau_s^2, \tag{23}$$

where $\tau_s \simeq l/v_{\rm rel}$.

A DM clump acquires the maximum energy gain during a single encounter with a star when impact parameter $l \sim R$. Integrating over all impact parameters $l > R$, we calculate an averaged rate of mass loss by a clump during successive encounters with stars:

$$\frac{1}{M}\left(\frac{dM}{dt}\right)_s \simeq \frac{1}{2T_t\sqrt{2\pi G\rho_0}} \int\limits_R^\infty 2\pi l\, dl \int\limits_{x_{\rm min}}^{x_{\rm max}} \frac{ds\, n_*(s) v_{rel}}{\sqrt{\varepsilon - \psi(s) - y/s^2}} \left(\frac{\delta M}{M}\right)_s, \tag{24}$$

where $n_*(r)$ is a radial number density distribution of stars in the bulge and halo.

4. Surviving fraction of clumps

From Eq. (11) it is seen that the tidal forces influence mainly the outer part of clump (where Δz is rather large). Further we use our basic assumption that only outer layers of a clump undergo the tidal stripping, while the inner parts of a clump are unaffected by tidal forces. Thus we assume that a clump mass $M = M(t)$ and radius $R = R(t)$ are both diminishing in time due to the tidal stripping of outer layers but its internal density profile remains the same as given by Eq. (12), e. g., for the isothermal density profile $M(t) \propto R(t)$ and $\bar{\rho}(t) \propto M(t)^{-2}$. Combining together the rates of mass loss (21) and (24) due to the tidal stripping of a clump by the disk and stars respectively, we obtain the evolution equation for a clump mass:

$$\frac{dM}{dt} = \left(\frac{dM}{dt}\right)_d + \left(\frac{dM}{dt}\right)_s. \tag{25}$$

In the following we solve this equation numerically starting from the time of Galaxy formation at $t_0 - t_G$ up to the present moment t_0.

The most important astrophysical manifestation of DM clumps is a possible annihilation of the constituent SUSY DM particles. The crucial point is a dominance of the central core of a clump in annihilation signal if clumps have a steep enough density profile. Namely, annihilation of DM particles in a clump core prevail in a total annihilation rate in a single clump with a power-law density profile (1) if $\beta > 3/2$ and $x_c = R_c/R \ll 1$. More specifically, the quantity $\dot{N} \propto \int_{r_0}^{r} 4\pi r'^2 dr' \rho_{\rm int}^2(r')$ practically does not depend on r, if $r \gg r_0$. As a result the annihilation luminosity of DM clump with approximately isothermal density profile ($\beta \simeq 2$) will be nearly constant under influence of tidal stripping until a clump radius diminishes to its core radius. In other words, in the nowadays Galaxy the remnants of tidally stripped clumps with $x_c < \mu(t_0) \ll 1$, where $\mu(t) = M(t)/M_i$ and $t_0 \simeq 10^{10}$ yrs is the Galaxy age, obeys the evolution equation (25) and have the same annihilation luminosity as their progenitors with $\mu = 1$.

By using evolution equation (25) we now calculate the probability P of the survival of clump remnant during the lifetime of the Galaxy. Let us choose some arbitrary point in the halo with a radius-vector $\vec{r}$ and an angle α with a polar axis of the Galactic disk. Only the clump orbits with inclination angle $\pi/2 - \alpha < \gamma < \pi/2$ pass through this point. A survival probability for clumps can be written now in the following form

$$P(x,\alpha) = \frac{4\pi\sqrt{2}}{\bar{\rho}(x)\sin\alpha} \int\limits_0^1 dp \int\limits_0^{\sin\alpha} d\cos\gamma \int\limits_{\psi(x)}^1 d\varepsilon\, [\varepsilon - \psi(x)]^{1/2} F(\varepsilon)\Theta[\mu(t_0) - x_c]. \tag{26}$$

In this equation $\tilde{\rho}(x)$ is a density profile of the halo from (19), $p = \cos\theta$, θ is an angle between the radius-vector $\vec{r}$ and the orbital velocity of clump, Θ is the Heaviside function, $\psi(x)$ is the halo gravitational potential from (19), $F(\varepsilon)$ is a distribution function of clumps in the halo, $\mu(t_0)$ depends on all variables of the integration and $x_c = R_c/R$ is an initial value of clump core. The function $\mu(t_0)$ is calculated from numerical solution of evolution equation (25). If $\mu(t_0) > x_c$, the clump remnant is survived through the tidal destruction by both the disk and stars. The annihilation rate in this remnant would be the same as in the initial clump. On the contrary in the opposite case, when $\mu(t_G) < x_c$, the clump is totally destructed because (i) the core is not a dynamically separated system and composed of particles with extended orbits, and because (ii) a nearly homogeneous core is destructed easier than a similar object with the same mass but with a near isothermal density profile.

According to theoretical model [3] and numerical simulations [29], a differential number density of small-scale clumps in the co-moving frame in the Universe is $n(M)\,dM \propto dM/M^2$. This distribution is calculated in [6]. The damping of small-scale perturbations with $M < M_{\rm min}$ provides an additional factor $\exp[-(M/M_{\rm min})^{2/3}]$ responsible for the fading of distribution at small M. The result of numerical simulations [29] can be expressed in the form of a differential mass fraction of the DM clumps in the Galactic halo $f(M)\,dM \simeq \kappa(dM/M)$, where $\kappa \simeq 8.3 \times 10^{-3}$. The analytical estimation (10) gives approximately $\kappa \simeq 4 \times 10^{-3}$ for the mass interval $10^{-6}{\rm M}_\odot < M < 1{\rm M}_\odot$. The discrepancy by the factor $\simeq 2$ may be attributed to the approximate nature of our approach as well as to the well known additional factor 2 in the original Press–Schechter derivation of the mass function. In the latter case one must simply multiply equation (7) by factor 2.

5. Amplification of annihilation signal

A local annihilation rate is proportional to the square of DM particle number density. A number density of DM particles in clump is much large than a corresponding number density of the diffuse (not clumped) component of DM. For this reason an annihilation signal from even a small fraction of DM clumps can dominate over an annihilation signal from the diffuse component of DM in the halo. We consider here the Harrison-Zeldovich initial perturbation spectrum with power index $n_p = 1$ as a representative example. The value of n_p is not exactly fixed by the current observations of CMB anisotropy. In the case of $n_p < 1$, the DM clumps are less dense, and a corresponding amplification of annihilation signal would be rather small [3].

The gamma-ray flux from annihilation of diffuse distribution (19) of DM in the halo is proportional to

$$I_{\rm H} = \int\limits_0^{r_{\rm max}(\zeta)} \rho_H^2(\xi)\,dx, \tag{27}$$

where the integration is over r goes along the line of sight, $\xi(\zeta, r) = (r^2 + r_\odot^2 - 2rr_\odot\cos\zeta)^{1/2}$ is the distance to the Galactic center, $r_{\rm max}(\zeta) = (R_{\rm H}^2 - r_\odot^2\sin^2\zeta)^{1/2} + r_\odot\cos\zeta$ is a distance to the external halo border, ζ is an angle between the line of observation and the direction to the Galactic center, $R_{\rm H}$ is a virial radius of the Galactic halo, $r_\odot = 8.5$ kpc is the distance between the Sun and Galactic center. The corresponding signal from annihilations of DM in clumps is proportional to the quantity [3]

$$I_{\rm cl} = S \int\limits_0^{r_{\rm max}(\zeta)} dx \int\limits_{M_{\rm min}} f(M)\,dM \rho\rho_H(\xi)P(\xi,\rho), \tag{28}$$

where $\rho(M)$ is the mean density of clump. The function S depends on the clump density profile and core radius of clump [3] and we use $S \simeq 14.5$ as a representative example. The observed

Journal of Physics: Conference Series **607** (2015) 012015 doi:10.1088/1742-6596/607/1/012015

amplification of the annihilation signal is defined as $\eta(\zeta)=(I_{\rm cl}+I_{\rm H})/I_{\rm H}$ is calculated numerically in [3, 6] for the case $x_c=0.1$. It tends to unity at $\zeta\to 0$ because of the divergent form of the halo profile (19). The annihilation of diffuse DM prevails over signal from clumps at the the Galactic center. The observed signal is obtained by integration along the line of sight and the effect of clumps destruction at the Galactic center is masked by the signal from another regions of the halo. This amplification of an annihilation signal is called a "boost-factor" [47].

6. Anisotropy of the observed signal

The usual assumption in calculations of DM annihilation is a spherical symmetry of the Galactic halo. In this case an anisotropy of annihilation gamma-radiation is only due to off-center position of the Sun in the Galaxy. Nevertheless, a principal significance of the halo nonsphericity for the observed annihilation signal was demonstrated in [48]. According to observations, the axes of the Galactic halo ellipsoid differ most probably no more than $10-20\%$, but even a much more larger difference of axes, up to a factor 2, can not be excluded [49, 50]. This leads to more than an order of magnitude uncertainty in the predicted annihilation flux from the Galactic anti-center direction [48]. It was shown [3, 6] that (i) small-scale DM clumps dominate in the generation of annihilation signal and (ii) the Galactic stellar disk provides the main contribution to the tidal destruction of clumps at $r>3$ kpc, i.e., outside the central bulge region. The detectors at the Fermi satellite are sensitive to anisotropy up to 0.1% level [51]. This provide a hope to discriminate the anisotropic DM annihilation signal from the diffuse gamma-ray backgrounds. To calculate the anisotropy we used a simplified criterium for a tidal destruction of clump: clump is destructed if a total tidal energy gain $\sum(\Delta E)_j$ after several tidal shocking becomes of order of initial binding energy of a clump $|E|$, i.e., $\Delta E=\sum\limits_j(\Delta E)_j\sim|E|$, where summation goes over the successive disk crossings (or encounters with stars). In this model the survival probability for clumps can be written now in the following form

$$P(r,\alpha)=\frac{4\pi\sqrt{2}}{\tilde{\rho}(x)\sin\alpha}\int\limits_0^1 dp\int\limits_0^{\sin\alpha}d\cos\gamma\int\limits_{\psi(x)}^1 d\varepsilon\,[\varepsilon-\psi(x)]^{1/2}F(\varepsilon)\,e^{-\Delta E/|E|}, \tag{29}$$

where the definitions are the same as in (26) except for exponential factor for clumps destruction $e^{-\Delta E/|E|}$. In [3] the triple integral from (29) for survival probability $P(r,\alpha)$ was numerically calculated. As a representative example we consider the Earth-mass clumps $M=10^{-6}M_\odot$ originated from 2σ density peaks in the case of power-law index of primordial spectrum of perturbations $n_p=1$. The mean internal density of these clumps is $\rho\simeq 7\times10^{-23}$ g cm^{-3}. The annihilation signal for the Galactic disk plane and for the orthogonal vertical plane (passing through the Galactic center) as a function of angle ζ between the observation direction and the direction to the Galactic center was calculated numerically in [11]. The difference of the signals in two orthogonal planes at the same ζ can be considered as an anisotropy measure. Defined as $\delta=(I_2-I_1)/I_1$, it has a maximum value $\delta\simeq 0.09$ at $\zeta\simeq 39°$.

7. Conclusion

We demonstrate that the cores of the small-scale DM clumps (or clump remnants) survive in general during clump formation and their tidal destruction by stars in the Galaxy. These small-scale DM clumps may provide the major contribution to the annihilation signal (in comparison with the diffuse DM) in the Galactic halo. The amplification (boost-factor) can reach 10^2 or even 10^3 depending on the initial perturbation spectrum and minimum mass of clumps.

References

[1] Schmid C, Schwarz D J and Widerin P 1999 *Phys. Rev.* D **59** 043517

[2] Schwarz D J and Hofmann S 2000 *Nucl. Phys. Proc. Suppl* **87** 93
[3] Berezinsky V, Dokuchaev V and Eroshenko Yu 2003 *Phys. Rev.* D **68** 103003
[4] Zhao H S, Taylor J, Silk J and Hooper D arXiv:astro-ph/0502049.
[5] Green A M, Hofmann S and Schwarz D J 2005 *JCAP* **0508** 003
[6] Berezinsky V, Dokuchaev V and Eroshenko Yu 2006 *Phys. Rev.* D **73** 063504
[7] Diemand J, Kuhlen M and Madau P 2006 *Astrophys. J.* **649**
[8] Green A M and Goodwin S P 2007 *Mon. Not. Roy. Astron. Soc.* **375** 1111
[9] Bertschinger E 2006 *Phys. Rev.* D **74** 063509
[10] Angus G W and Zhao H S 2007 *Mon. Not. Roy. Astron. Soc.* **375** 1146
[11] Berezinsky V, Dokuchaev V and Eroshenko Yu *JCAP* **07** 011 (2007).
[12] C. Giocoli, L. Pieri and G. Tormen, *Mon. Not. Roy. Astron. Soc.* **387** 689-697 (2008).
[13] Berezinsky V, Dokuchaev V, Eroshenko Yu, Kachelriess M and Solberg M Aa 2010 *Phys. Rev.* D **81** 103529
[14] Berezinsky V, Dokuchaev V, Eroshenko Yu, Kachelriess M and Solberg M Aa 2010 *Phys. Rev.* D **81** 103530
[15] Berezinsky V, Dokuchaev V, Eroshenko Yu 2011 *JCAP* **12** 007
[16] Berezinsky V, Dokuchaev V, Eroshenko Yu 2013 *JCAP* **11** 059
[17] Weinberg S 1971 *Astrophys. J.* **168** 175
[18] Hofmann S, Schwarz D J and Stocker H 2001 *Phys. Rev.* D **64** 083507
[19] Loeb A and Zaldarriaga M 2005 *Phys. Rev.* D **71** 103520
[20] Silk J 1968 *Astrophys. J.* **151** 459
[21] Zybin K P, Vysotsky M I and Gurevich A. V 1999 *Phys. Lett.* A **260** 262
[22] Kamionkowsky M and Koushiappas S M 2008 *Phys. Rev.* D **77** 103509
[23] Pieri L, Bertone G and Branchini E 2008 *Mon. Not. Roy. Astron. Soc.* **384** 1627
[24] Ando S and Komatsu E 2006 *Phys. Rev.* D **73** 023521
[25] Myers Z and Nusser A 2008 *Mon. Not. Roy. Astron. Soc.* **384** 727
[26] Gurevich A V and Zybin K P 1988 *JETP* **67** 1
[27] Gurevich A V and Zybin K P 1988 *JETP* **67** 1957 (1988)
[28] Gurevich A V and Zybin K P 1995 *Phys. Uspekhi* **38** 687 (1995).
[29] Diemand J, Moore B and Stadel J 2005 *Nature* **433** 389
[30] Doroshkevich A G, Lukash V N and Mikheeva E V 2007 **Nuovo Cim.** B **122** 1393
[31] Kaplinghat M 2005 *Phys. Rev.* D **72** 063510
[32] Strigari L E, Kaplinghat M, and Bullock J S 2007 *Phys. Rev.* D **75** 061303
[33] Gnedin O Y and Ostriker J.P 1999 *Astrophys. J.* **513** 626
[34] Taylor J E and Babul A 2001 *Astrophys. J.* **559** 716
[35] Berezinsky V, Dokuchaev V and Eroshenko Yu 2008 *Phys. Rev.* D **77** 083519
[36] Diemand J, Kuhlen M and Madau P 2007 *Astrophys. J.* **667** 859
[37] Goerdt T, Gnedin O Y, Moore B, Diemand J and Stadel J 2007 *Mon. Not. Roy. Astron. Soc.* **375** 191
[38] Lacey C and Cole S 1993 *Mon. Not. Roy. Astron. Soc.* **262** 627
[39] Gnedin O Y, Hernquist L and Ostriker J P 1999 *Astrophys. J.* **514** 109
[40] Spitzer L Jr 1987 *Dynamical Evolution of Globular Clusters* (Princeton Univ. Press) Princeton
[41] Bugaev E V and Konishchev K V 2002 *Phys. Rev.* D **65** 123005
[42] Spergel D N et al 2003 *Astrophys. J.* Suppl. **148** 175
[43] Ostriker J P, Spitzer L Jr and Chevalier R A 1972 *Astrophys. J.* Lett. **176** 51
[44] Weinberg M. D. 1994 *Astron. J* **108** 1403
[45] Eddington A S 1916 *Mon. Not. Roy. Astron. Soc.* **76** 572
[46] Widrow L M 2000 *Astrophys. J.* Supp. **131** 39
[47] de Boer W, Sander C, Zhukov V, Gladyshev A V and Kazakov D.I 2005 *Astron. Astrophys.* **444** 51
[48] Calcaneo-Roldan C and Moore B 2000 *Phys. Rev.* D **62** 123005
[49] Olling R P and Merrifield M R 2000 *Mon. Not. Roy. Astron. Soc.* **311** 361
[50] Olling R P and Merrifield M R 2001 *Mon. Not. Roy. Astron. Soc.* **326** 164
[51] Hooper D and Serpico P D 2007 *JCAP* **06** 013
[52] Battaglia M, De Roeck A, Ellis J, Gianotti F, Olive K A and Pape L 2004 *Eur. Phys. J.* **C33** 273

In memory of Mikhail Igorevich Polikarpov

V. B. Bornyakov

Institute for High Energy Physics, 124281 Protvino, Russia; Institute of Theoretical and Experimental Physics, 117259 Moscow, Russia

P. V. Buividovich

Institute for Theoretical Physics, Regensburg University, Universitatsstrasse 31, D-93053 Regensburg, Germany

M.A. Zubkov

University of Western Ontario, Department of Applied Mathematics, London, ON, Canada N6A 5B7;
ITEP, B.Cheremushkinskaya 25, Moscow, 117259, Russia

Abstract. This obituary is devoted to M. I. Polikarpov (28.12.1952 - 18.07.2013). We recollect some facts of his biography and stages of his scientific career, and make a brief review of some of his most known scientific works.

The first published paper [1] by Mikhail Polikarpov is dated 1975. It was devoted to the general properties of amplitudes in hadronic physics. Within the period between 1975 and 1982 more than 20 papers were published on the similar subjects in collaboration with Yu. A. Simonov and A. M. Badalyan. The review of the results obtained during this period appeared in Physics Reports [2] in 1982.

Mikhail started to study lattice gauge theories in collaboration with Yu. M. Makeenko and A. I. Veselov [3, 4]. In particular, in [4] the results of the first numerical investigation of instantons in gluodynamics were presented. The properties of gauge field configurations obtained by cooling the quantum fluctuations of the gauge fields on the lattice were investigated. It was found that the cooling process can be subdivided into three distinct stages: the emergence of semi-classical instanton-anti-instanton vacuum, annihilation of instantons and anti-instantons, and , finally, relaxation to stable classical solutions. The distributions of instanton radii and inter-instanton distances were obtained. It was demonstrated that the contribution of the multi-instanton configurations to the value of the string tension was about 5%.

Remarkably, the analytical multi-instanton solution was also obtained by M. I. Polikarpov for the first time. This result remained unpublished and we became aware of it only after his death. A. A. Belavin (who was one of the authors of the first paper on instantons [5]) told us this story. The authors of [6] derived the equations that define the multi-instanton solutions in the Yang-Mills theory. M. I. Polikarpov was the first who obtained the analytical solutions of these equations. This was mentioned in [6], where these solutions were presented. It is worth mentioning, that independently and simultaneously the same problem was solved by E. Witten [7].

28.12.1952 - 18.07.2013

Prof. M. Polikarpov started to create his own scientific group in 1990. The first paper published by this group [8] was written in collaboration with his students T. L. Ivanenko and A. V. Pochinsky. In this paper it was numerically demonstrated in $SU(2)$ lattice gauge theory in maximal Abelian gauge that the density of Abelian monopoles which occupy several lattice cells (extended monopoles) is strongly correlated with the string tension. This result indicates that the string between quark and antiquark is the dual analog of the Abrikosov vortex in superconductors. In this study for the first time the Hausdorf dimension of the monopole current line was used as an order parameter.

After M. I. Polikarpov received his Habilitation degree in 1991, one of us (M. A. Z.) became his student. In 1993 the paper [9] was written, in which the partition function of the 4D lattice Abelian Higgs theory is represented as the sum over world sheets of Nielsen-Olesen strings. The creation and annihilation operators of these strings were constructed. The topological long-range interaction of the strings and charged particles was shown to exist; it is proportional to the linking number of the string world sheet and particle world trajectory.

In the later years Emil Akhmedov, Maxim Chernodub, and Fedor Gubarev joined the group of Prof. Polikarpov. As well as M. A. Z. they became the students of M. I. Polikarpov. Some of the most known works done in collaboration with Fedor and Maxim are [11, 12], where the authors have performed numerical calculation of the probability distribution of the value of the monopole creation operator in $SU(2)$ lattice gauge theory. It was found that in the low-temperature confining phase the maximum of this distribution is shifted from zero, which means that the effective constraint potential is of the Higgs type. Above the phase transition the minimum of the potential (the maximum of the monopole field distribution) is at the zero value of the monopole field. This fact confirms the existence of the Abelian monopole condensate in the confinement phase of lattice gluodynamics, and agrees with the dual superconductor model of the confining vacuum.

Emil Akhmedov after getting his PhD almost immediately became an independent researcher. Nevertheless, his own line of research began from the work [13] written under the supervision of M. I. Polikarpov. In this paper the continuum theory of quantum Abrikosov-Nielsen-Olesen strings was constructed starting from the Abelian Higgs field theory. It was shown that in four space-time dimensions in the limit of infinitely thin strings, the conformal anomaly is absent, and the quantum theory of such strings exists.

The new impulse was given to the ITEP lattice group when it was joined by the world famous physicist V. I. Zakharov. His experience and scientific intuition became the important ingredient of the further research work carried out by the group. With his participation, for example, the paper on anatomy of Abelian monopoles appeared [14]. In this paper the Abelian and non-Abelian action densities near the monopole in the maximal Abelian gauge of SU(2) lattice gauge theory were studied. It was found that the non-Abelian action density near the monopoles belonging to the percolating cluster decreases when the monopole center is approached. The estimate of the monopole radius was given (≈ 0.04 fm). Approximately at that time one of us (V. G. B.) joined ITEP lattice group as well.

The next generation of the members of the next generation of the ITEP lattice group are A. V. Kovalenko, S. N. Syritsyn, S. M. Morozov and P. Yu. Boyko, who were the students at ITEP. Later S. N. Syritsyn moved to US, while A. V. Kovalenko, S. M. Morozov and P. Yu. Boyko became PhD students and then researchers at ITEP. About at that time V.A. Belavin became the PhD student of M.I.Polikarpov. After getting PhD he moved to the investigation of conformal field theories. In the series of papers written by M. I. Polikarpov in collaboration with V. G. Bornyakov, M. N. Chernodub, F. V. Gubarev, V. I. Zakharov, A. V. Kovalenko, S. N. Syritsyn, S. M. Morozov and P. Yu. Boyko the lattice studies of the structure of QCD vacuum were continued. Some of the well-known publications of this period are [15, 16]. In [15] measurements of the action associated with center vortices in $SU(2)$ pure lattice gauge theory were performed. In lattice units the excess of the action on the plaquettes belonging to the vortex turned out to be approximately constant and independent on the lattice spacing a. Therefore, the action of the center vortex is of order A/a^2, where A is its area. Since the area A is known to scale in the physical units, the measurements imply that the suppression due to the surface action is balanced, or fine tuned to the entropy factor which is to be an exponential of A/a^2.

In [16] propagators of the diagonal and the off-diagonal gluons were studied numerically in the maximal Abelian gauge of $SU(2)$ lattice gauge theory. It was found that in the infrared region the propagator of the diagonal gluon was strongly enhanced in comparison with the off-diagonal one. The enhancement factor is about 50 at the smallest momentum 325 MeV.

Around 2007 one of us (P. V. B.) and Elena Luschevskaya joined the ITEP lattice group. Both were the PhD students of M. I. Polikarpov, and then researchers at ITEP. At this time the research direction of the group shifted to the lattice studies of the effect of strong magnetic fields on QCD vacuum. In particular, in [17] the microscopic picture of the Chiral Magnetic Effect was studied on the lattice. There is a recent evidence that this effect is observed by the STAR Collaboration in heavy ion collisions at RHIC. In [17] it was found that indeed magnetic field strongly enhances the fluctuations of electric current in the direction of the magnetic field. In [18] this enhancement was related to the emergence of electric conductivity in the direction of magnetic field. It was predicted that this phenomenon should manifest itself in specific anisotropy of the distribution of soft leptons produced in heavy-ion collisions.

The last scientific topic investigated by Misha Polikarpov was the physics of graphene. In [19] a first-principle numerical study of spontaneous breaking of chiral (sublattice) symmetry in suspended monolayer graphene due to electrostatic interaction was performed. For the first time the screening of Coulomb potential by electrons on σ-orbitals was taken into account. In contrast to the results of previous numerical simulations with unscreened potential, it was found that suspended graphene is in the conducting phase with unbroken chiral symmetry. This finding is in agreement with the most recent experimental results obtained by the group of K. Novoselov and A. Geim.

M. I. Polikarpov collaborated fruitfully with the foreign colleagues from many countries. The

Journal of Physics: Conference Series **607** (2015) 012016 doi:10.1088/1742-6596/607/1/012016

very long collaborations were with T. Suzuki, M. Müller-Preussker, G. Schierholz, M. Ilgenfritz, G. Greensite, B. Bakker.

To conclude, let us simply list the former and current members of the ITEP lattice group, which was created by Misha Polikarpov, and which remains the only lattice group in Russia up to now: Emil Akhmedov, Vladimir Belavin, Vitaly Bornyakov, Pavel Boyko, Victor Braguta, Pavel Buividovich, Maxim Chernodub, Fedor Gubarev, Taras Ivanenko, Oleg Kochetkov, Anton Kononenko, Andrey Kotov, Alexey Kovalenko, Olga Larina, Elena Luschevskaya, Yurij Makeenko, Boris Martemyanov, Valentin Mitrjushkin, Alexander Molochkov, Sergey Morozov, Oleg Pavlovsky, Andrey Pochinsky, Mikhail Prokudin, Roman Rogalev, Vladimir Shevchenko, Sergey Syritsyn, Maxim Ulybyshev, Semen Valgushev, Alexander Veselov, Valentin Zakharov, Mikhail Zubkov.

The people listed above are the former students of Misha Polikarpov, his colleagues and friends. Not all members of the lattice ITEP group continue now to work in science. But, no doubts, all of them experienced the unique atmosphere of the scientific group created with love and passion by Misha Polikarpov and all of them will keep him in their hearts and memories forever.

[1] A. M. Badalian, M. I. Polikarpov and Y. A. Simonov, preprint ITEP-38-1975.
[2] A. M. Badalian, L. P. Kok, M. I. Polikarpov and Y. A. Simonov, Phys. Rept. **82** (1982) 31.
[3] Y. M. Makeenko and M. I. Polikarpov, Nucl. Phys. B **205** (1982) 386.
[4] M. I. Polikarpov and A. I. Veselov, Nucl. Phys. B **297** (1988) 34.
[5] A. A. Belavin, A. M. Polyakov, A. S. Schwartz and Y. S. Tyupkin, Phys. Lett. B **59** (1975) 85.
[6] D. E. Burlankov and V. N. Dutyshev, "Higher Order Instantons" (In Russian), Zh.Eksp.Teor.Fiz.**73** (1977) 377-381.
[7] E. Witten, "Some Exact Multi-Instanton Solutions of Classical Yang-Mills Theory", Phys.Rev.Lett. **38** (1977) 121.
[8] T. L. Ivanenko, A. V. Pochinsky and M. I. Polikarpov, Phys. Lett. B **252** (1990) 631.
[9] M. I. Polikarpov, U. J. Wiese and M. A. Zubkov, Phys. Lett. B **309** (1993) 133 [hep-lat/9303007].
[10] M. I. Polikarpov, Nucl. Phys. Proc. Suppl. **53** (1997) 134 [hep-lat/9609020].
[11] M. N. Chernodub, M. I. Polikarpov and A. I. Veselov, Phys. Lett. B **399** (1997) 267 [hep-lat/9610007].
[12] M. N. Chernodub, F. V. Gubarev, M. I. Polikarpov and A. I. Veselov, Prog. Theor. Phys. Suppl. **131** (1998) 309 [hep-lat/9802036].
[13] E. T. Akhmedov, M. N. Chernodub, M. I. Polikarpov and M. A. Zubkov, Phys. Rev. D **53** (1996) 2087 [hep-th/9505070].
[14] V. G. Bornyakov, M. N. Chernodub, F. V. Gubarev, M. I. Polikarpov, T. Suzuki, A. I. Veselov and V. I. Zakharov, Phys. Lett. B **537** (2002) 291 [hep-lat/0103032].
[15] F. V. Gubarev, A. V. Kovalenko, M. I. Polikarpov, S. N. Syritsyn and V. I. Zakharov, Phys. Lett. B **574** (2003) 136 [hep-lat/0212003].
[16] V. G. Bornyakov, M. N. Chernodub, F. V. Gubarev, S. M. Morozov and M. I. Polikarpov, Phys. Lett. B **559** (2003) 214 [hep-lat/0302002].
[17] P. V. Buividovich, M. N. Chernodub, E. V. Luschevskaya and M. I. Polikarpov, Phys. Rev. D **80** (2009) 054503 [arXiv:0907.0494 [hep-lat]].
[18] P. V. Buividovich, M. N. Chernodub, D. E. Kharzeev, T. Kalaydzhyan, E. V. Luschevskaya and M. I. Polikarpov, Phys. Rev. Lett. **105** (2010) 132001 [arXiv:1003.2180 [hep-lat]].
[19] M. V. Ulybyshev, P. V. Buividovich, M. I. Katsnelson and M. I. Polikarpov, Phys.Rev.Lett. **111** (2013) 056801 [arXiv:1304.3660 [cond-mat.str-el]].

Interaction of static charges in graphene

V V Braguta[1,2], S N Valgushev[2,3], A A Nikolaev[4], M I Polikarpov[2,3] and M V Ulybyshev[2,5]

[1] Institute for High Energy Physics, Protvino, 142281 Russia
[2] Institute of Theoretical and Experimental Physics, Moscow, 117218 Russia
[3] Moscow Institute for Physics and Technology, Dolgoprudny, 141700 Russia
[4] School of Biomedicine, Far Eastern Federal University, Vladivostok, 690950 Russia
[5] Institute for Theoretical Problems of Microphysics, Moscow State University, Moscow, 119899 Russia

E-mail: victor.v.braguta@gmail.com

Abstract. The study of the interaction potential between static charges within Monte-Carlo simulation of graphene is carried out. The numerical simulations are performed in the effective lattice field theory with noncompact 3 + 1-dimensional Abelian lattice gauge fields and 2 + 1-dimensional staggered lattice fermions. At low temperature the interaction can be described by the Coulomb potential reduced by some dielectric permittivity ϵ_R. The dependence of ϵ_R on the dielectric permittivity of substrate is determined. In addition, the renormalization of the quasiparticle charge is studied. At large temperatures the interaction potential can be described by the two-dimensional Debye screening. The dependence of Debye screening mass on the dielectric permittivity of substrate allows to determine the position of the insulator-semimetal phase transition. It is shown that graphene reveals in the semimetal phase the properties of the two-dimensional plasma of fermions excitations.

1. Introduction

Graphene is an allotrope of carbon, in which atoms form a two-dimensional honeycomb lattice. Carbon atoms in graphene are bonded by sp^3-bonds and the bond length is about 0.142 nm [1].

The charge carriers in graphene at low energies behave as massless Dirac fermions [2]. The Fermi velocity of charge carriers $v_F \approx c/300$. Since the Fermi velocity is much smaller than the speed of light, magnetic effects and effects of delay in the interactions between charge carriers in graphene may be neglected, thus electron-electron interaction in graphene is described by the instantaneous Coulomb potential. The effective coupling constant for the Coulomb interaction in suspended graphene $g^2 = e^2/v_F \approx 2$ is large, and this material may be considered as a strongly interacting system. For graphene on substrate with the dielectric permittivity ϵ the effective coupling constant is reduced by a factor $2/(\epsilon + 1)$. The variation of the dielectric permittivity ϵ of substrate changes effective coupling constant and thus allows to study the properties of graphene in both strong and weak coupling regime.

There exists a number of papers where graphene was studied by Monte-Carlo method [3, 4, 5, 6, 7] and insulator-semimetal phase transition was found. In the weak coupling regime ($\epsilon > 5$, $g^2 < 0.7$) graphene is in the semimetal phase. In this phase the conductivity is $\sigma \approx e^2/h$ and there is no gap in the spectrum of fermionic excitations. The chiral symmetry of graphene is not broken. In the strong coupling regime ($\epsilon < 3$, $g^2 > 1$) graphene is in the insulator phase. In this phase the conductivity is considerably suppressed and fermionic excitations acquire dynamical mass [8]. The phase transition between the weak and the strong coupling regimes takes place at the value of the dielectric permittivity of substrate $\epsilon \approx 4$ [3, 4, 6].

2. Lattice simulation of graphene

We present the results of MC simulations of graphene in the framework of effective field model. Potential between static charges in graphene plane was measured. The non-MC calculations of the potential were performed in [9] (see also references therein).

The partition function of the effective field theory of graphene can be written as [2, 10, 11, 12]:

$$\mathcal{Z} = \int \mathcal{D}\bar{\psi}\mathcal{D}\psi\mathcal{D}A_0 \exp\left(-\frac{1}{2}\int d^4x \left(\partial_i A_0\right)^2 - \int d^3x\, \bar{\psi}_f \left(\gamma_0\left(\partial_0 - igA_0\right) - \sum_{i=1,2}\gamma_i\partial_i\right)\psi_f\right), \quad (1)$$

where A_0 is the zero component of the vector potential of the $3+1$ D electromagnetic field, γ_μ are Euclidean gamma-matrices, ψ_f $(f = 1, 2)$ are two flavors of Dirac fermions which correspond to the two spin components of the non-relativistic electrons in graphene and effective coupling constant $g^2 = 2e^2/(v_F(\epsilon+1))$ ($\hbar = c = 1$ is assumed). The following notations will be used below:

$$\alpha_0 = e^2\frac{2}{\epsilon+1}, \; \alpha_R = \frac{\alpha_0}{\epsilon_R}, \quad (2)$$

where α_0 is the bare effective squared charge, α_R is the effective squared charge, renormalized due to interaction and ϵ_R is the effective dielectric permittivity of graphene.

To measure the potential, $V(\vec{r})$, between static charges in graphene, we calculate the correlator of two Polyakov lines $\langle P^\gamma(0)(P^\gamma(\vec{r}))^+\rangle$:

$$\langle P^\gamma(0)(P^\gamma(\vec{r}))^+\rangle = a\exp\left(-\frac{V(\vec{r})}{T}\right), \quad (3)$$

where $\vec{r}$ is restricted to the graphene plane, a parameterizes the self-energy part of the correlator and the Polyakov line $P(\vec{r})$ is

$$P(\vec{r}) = \exp\left(-ie\int_0^{1/T} dtA_0(t,\vec{r})\right) = \prod_{t=0}^{L_t-1}\exp\left(-i\theta_{(t,\vec{r}),0}\right). \quad (4)$$

To suppress statistical errors, we measure the rational power of the correlator of Polyakov lines. Physically this means that the interaction potential between static charges $\pm e\cdot\gamma$ is considered. We have found that for $\gamma \sim 0.1$ the uncertainty of the calculation is much smaller than for $\gamma = 1$ (usual Polyakov line). In the calculations the value $\gamma = 0.1$ was used. Technical details of the lattice regularization scheme and simulation algorithm may be found in [13].

The results of the numerical simulations of $V(\vec{r})$ in the correlator (3) is fitted by the lattice potential with screening

$$V(\vec{r}) \;=\; \frac{1}{\epsilon_R}V_{fit}(\vec{r}) + c \quad (5)$$

As a result the value of ϵ_R is determined. The constant c in formula (5) parameterizes the self-energy contribution to the potential. Two different potentials $V_{fit}(\vec{r})$ were used for fitting: Coulomb potential

$$V_C(\vec{r}) \;=\; \alpha_0\frac{\pi\gamma^2}{L_s^3 a_s}\sum_{n_1,n_2,n_3}\frac{1}{\sum_i \sin^2(p_i a_s/2)}e^{i\vec{p}\vec{r}} \quad (6)$$

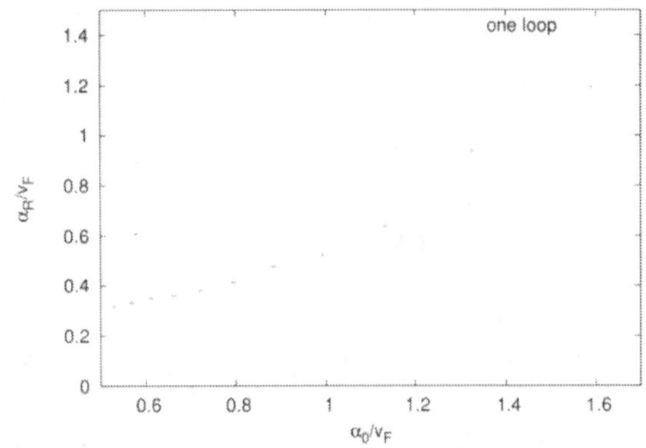

Figure 1. The renormalized squared charge α_R as a function of the bare squared charge α_0 (rescaled by the v_F) and the plot of one loop renormalization formula (8). The insulator-semimetal phase transition takes place at $\alpha_0/v_F \sim 0.9$.

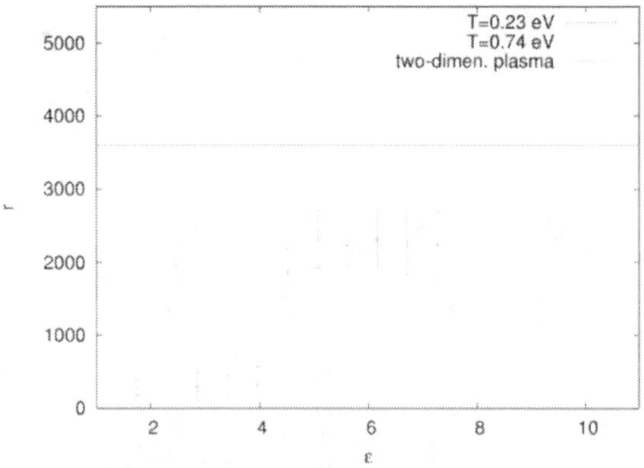

Figure 2. The dependence of the ratio $r = (m_D e^2)/(T\alpha_R)$ on the ϵ at different temperatures. The line parallel to the ϵ-axis is the value of the ratio $(m_D e^2)/(T\alpha_R)$ at the approximation of weakly interacting two-dimensional plasma of quasiparticles.

and Debye potential

$$V_D(\vec{r}) = 4\pi\alpha_0\gamma^2 \sum_{n_1,n_2} \frac{f(p_1,p_2)}{1+2m_D(L_s a_s)^2 f(p_1,p_2)} e^{i\vec{p}\vec{r}}, \; f(p_1,p_2) = \frac{1}{4L_s^3 a_s} \sum_{n_3} \frac{1}{\sum_i \sin^2(p_i a_s/2)}, \quad (7)$$

where in both formulae $p_i = (2\pi n_i)/(L_s a_s)$ and the integers n_1, n_2, n_3 run over the values $0, 1, .., L_s - 1$ (the point $n_1 = n_2 = n_3 = 0$ is excluded). Debye potential without explicit expression for the screening mass can be thought of as a modification of the Coulomb potential with unknown parameter m_D (at $m_D = 0$ formula (7) evolves into (6)).

3. Numerical results and discussion

First, let us consider the dependence of ϵ_R on ϵ at low temperature. In Fig. 1 we show how α_R is renormalized due to the interaction ($T = 0.23$ eV). In the semimetal phase the effective coupling constant is not large ($\alpha_0/v_F < 1$) and one can try to apply perturbation theory to desribe our data. In the one loop approximation the dependence of α_R on the α_0 for graphene is given by the following expression [14]:

$$\frac{\alpha_R}{\alpha_0} = \frac{1}{1+\frac{\pi}{2}\frac{\alpha_0}{v_F}} = \frac{1}{1+3.4\frac{2}{\epsilon+1}}. \quad (8)$$

Fig. 1 shows that at small α_0 we have a good agreement with perturbation theory.

Now let us turn to Debye screening mass. Evidently, m_D is equal to zero if there is no interaction between quasiparticles. This means that the expansion of m_D starts from the term proportional to α_R, which determines the strength of the interaction. The second property of the m_D is its disappearance in the case of zero quasiparticles density n. One concludes that $m_D \sim (\alpha_R n)/T$. In Fig. 2 we present the observable $r = (m_D e^2)/(T\alpha_R)$, which is proportional to n/T^2. This observable allows to study the density of quasiparticles in graphene. If the interaction between the quasiparticles is weak, the ratio $(m_D e^2)/(T\alpha_R)$ is equal to $r = 8\log 2\frac{e^2}{v_F^2} \simeq 3600$ [13].

One can see from Fig. 2 that *at low temperature Debye mass m_D plays the role of order parameter for the insulator-semimetal phase transition.* At small values of the dielectric permittivity of substrate m_D is equal to zero within the accuracy of the calculation, this means that the interaction potential is the ordinary Coulomb potential. At $\epsilon \sim 4-5$ Debye mass becomes nonzero, abruptly reaching the regime of the two-dimesional plasma. Thus the study of Debye screening mass allows to determine the position of the insulator-semimetal phase transition, which takes place at $\epsilon \sim 4-5$, in accordance with the results of papers [3, 4, 6]. At large temperatures m_D is nonzero for any values of the ϵ. This is a smoothly increased function of ϵ, which is saturated at $\epsilon \sim 4-5$.

Now let us consider the semimetal phase in the region $\epsilon > 5$. In this region the ratio $(m_D e^2)/(T\alpha_R)$ tends to a certain constant value and this value is $\sim 1.5-2.0$ times smaller than the one given by the theoretical predictions. The possible source of this disagreement is that we used the bare Fermi velocity v_F. The renormalized Fermi velocity v_F^R is larger than v_F, and taking into account v_F^R will push the value into the correct direction. The ratio v_F^R/v_F may be estimated as $\sim 1.2-1.4$ according to Fig. 2. This value is in a reasonable agreement with the results obtained within Monte-Carlo simulation of graphene [15] and with experiment [16]. Taking into account the renormalization of Fermi velocity one can conclude that *in the semimetal phase electron excitations in graphene form a weakly interacting two-dimensional plasma.*

Acknowledgments

The authors are grateful to Mikhail Zubkov for interesting and useful discussions. The work was supported by Grant RFBR-11-02-01227-a, by the Russian Ministry of Science and Education, under contract No. 07.514.12.4028 and by Development Program of Far Eastern Federal University. Numerical calculations were performed at the ITEP system Stakan (authors are much obliged to A.V. Barylov, A.A. Golubev, V.A. Kolosov, I.E. Korolko, M.M. Sokolov for the help), the MVS 100K at Moscow Joint Supercomputer Center and at Supercomputing Center of the Moscow State University.

References

[1] Heyrovska R 2008 Atomic Structures of Graphene, Benzene and Methane with Bond Lengths as Sums of the Single, Double and Resonance Bond Radii of Carbon *Preprint* arXiv:0804.4086
[2] Semenoff G W 1984 *Phys. Rev. Lett.* **53** 2449
[3] Drut J E and Lähde T A 2009 *Phys. Rev. Lett.* **102** 026802; 2009 *Phys. Rev.* B **79** 165425; 2009 *Phys. Rev.* B **79** 241405
[4] Drut J E, Lähde T A and Tölö E, 2010 *Proc. XXVIII Int. Symp. on Lattice Field Theory (Villasimius)* PoS(Lattice 2010)006; 2011 *Proc. XXIX Int. Symp. on Lattice Field Theory (Squaw Valley)* PoS(Lattice 2011)074
[5] Hands S and Strouthos C 2008 *Phys. Rev.* B **78** 165423; Armour W, Hands S and Strouthos C 2010 *Phys. Rev.* B **81** 125105; 2011 *Phys. Rev.* B **84** 075123
[6] Buividovich P V, Luschevskaya E V, Pavlovsky O V, Polikarpov M I and Ulybyshev M V 2012 *Phys. Rev.* B **86** 045107
[7] Buividovich P V and Polikarpov M I 2012 *Phys. Rev.* B **86** 245117
[8] Araki Y 2011 *Annals of Physics* **326** 1408–24
[9] M van Schilfgaarde and Katsnelson M I 2011 *Phys. Rev.* B **83** 081409
[10] Novoselov K S, Geim A K, Morozov S V, Jiang D, Zhang Y, Dubonos S V, Grigorieva I V and Firsov A A 2004 *Science* **306** 666
[11] Castro Neto A H, Guinea F, Peres N M R, Novoselov K S and Geim A K 2009 *Rev. Mod. Phys.* **81** 109
[12] Geim A K and Novoselov K S 2007 *Nature Materials* **6** 183
[13] Braguta V V, Valgushev S N, Nikolaev A A, Polikarpov M I and Ulybyshev M V 2013 Interaction of static charges in graphene within Monte-Carlo simulation *Preprint* arXiv:1306.2544
[14] Gonzalez J, Guinea F and Vozmediano M A H 1994 *Nucl. Phys.* B **424** 595
[15] Drut J E and Lähde T A 2013 Fermi velocity renormalization and the excitonic insulator in graphene *Preprint* arXiv:1304.1711
[16] Yu G L *et al* 2013 *Proc. National Academy of Sciences of the USA* **110**(9) 3282-86

Axial Magnetic Effect and Chiral Vortical Effect with free lattice chiral fermions

P. V. Buividovich

Institute for Theoretical Physics, Regensburg University, Universitätsstrasse 31, D-93053 Regensburg, Germany

E-mail: pavel.buividovich@physik.uni-regensburg.de

Abstract. Following the recent work by Braguta *et al.*, we perform an indirect calculation of the chiral vortical conductivity σ_V with free lattice overlap fermions by measuring the response of the energy flow to the constant axial magnetic field. We find that the measurements are contaminated by very large finite-volume artifacts. However, the continuum result for finite-temperature contribution to the chiral vortical conductivity $\sigma_V = \frac{T^2}{12}$ is still reproduced in the limit of infinite volume at fixed temperature.

Transport phenomena in chirally imbalanced medium have become a subject of intense research in recent years. These phenomena are closely related to quantum anomalies and for that reason are extremely interesting from theoretical point of view. It has been also argued that they might manifest themselves in heavy-ion collision experiments (see [1] for an extensive collection of reviews). Very recently, these arguments have been supported by fully nonlinear simulations of the anomalous hydrodynamics [2].

One of such anomalous transport phenomena is the Chiral Vortical Effect (CVE) - the generation of axial current $j_{A\,i}$ (where $i = x, y, z$ labels spatial coordinates) in the direction of the medium vorticity $\omega_i = \epsilon_{ijk}\partial_j v_k$, where v_i is the medium velocity [3, 4, 5]:

$$j_{A\,i} = \sigma_V\,\omega_i, \tag{1}$$

$$\sigma_V = \left(\frac{\mu_A^2 + \mu_V^2}{4\pi^2} + \frac{T^2}{12}\right) \tag{2}$$

where σ_V is the so-called Chiral Vortical Conductivity, μ_A and μ_V are the chiral and the baryon chemical potentials and T is the medium temperature. In the other words, vortices in a chiral liquid create chirality imbalance. Interestingly, this effect exists even in a neutral medium at finite temperature. It also turns out that the chiral vortical conductivity (2) might receive nontrivial contributions due to gauge interactions between fermions [6, 7].

In the linear response approximation, the chiral vortical conductivity σ_V is expressed in terms of the static correlator of the axial current density $j_{A\,i}$ and the energy flow vector T_{0i} [8]:

$$\sigma_V = \lim_{p_k \to 0} \frac{i\,\epsilon_{ijk}}{2p_k} \int dt d^3x \langle\, j_{A\,i}(x,t)\, T_{0j}(0,0)\,\rangle\, e^{ip_k x_k}, \tag{3}$$

where

$$T^{0i} = \frac{i}{2}\bar{\psi}\gamma_0\partial_i\psi + \frac{i}{2}\bar{\psi}\gamma_i\partial_0\psi \tag{4}$$

are the components of the fermionic energy-momentum tensor which are responsible for the energy flow in the i-th direction.

While it is conceptually very difficult to implement matter rotation on the lattice mainly due to the problems with periodic boundary conditions (see [9] for some work in this direction), it is rather easy to measure the correlator (3) on the lattice. One can simplify the problem even further by noticing that the expression (3) in fact also describes another anomalous transport phenomenon (sometimes termed Axial Magnetic Effect) - the generation of nonzero energy flow T_{0i} in the direction of the axial magnetic field $B_{A\,i} = \epsilon_{ijk}\partial_j A_k$. Here A_k is the axial gauge field which couples to Dirac fermions as $A_\mu \bar{\psi}\gamma_5\gamma_\mu\psi$. In close analogy with the lattice studies of the Chiral Magnetic Effect (see e.g. [10]) one can then introduce a constant axial magnetic field on the lattice and measure the energy flow along it. Technically, an advantage of the approach of [11] is that only the expectation value of a single fermionic bilinear operator should be measured, rather than the correlator of two operators in (3). This significantly simplifies algorithms and improves the signal-to-noise ratio. The results of such measurements in $SU(2)$ quenched lattice gauge theory with zero chiral and baryonic chemical potentials were reported recently in [11]. It was found that the chiral vortical conductivity is equal to zero within statistical errors in the confinement phase and is almost by an order of magnitude smaller than the result (2) in the deconfinement phase at the temperature $T = 1.58\,T_c$. However, one can expect that by virtue of asymptotic freedom in the limit of very high temperatures the chiral vortical conductivity approaches its value (2) for free fermions.

On the other hand, in the recent work [12] it was demonstrated that anomalous transport phenomena such as Chiral Magnetic Effect are extremely sensitive to finite volume and finite temperature effects. In particular, it was found that lattice measurements of CME with free chiral lattice fermions in the background of a constant magnetic field in a finite volume yield only half of the result which one can obtain from anomaly-based arguments. Preliminary studies have also indicated that the value of the chiral magnetic conductivity is also extremely sensitive to lattice artifacts which lead to the non-conservation of the vector current. Such artifacts are also present in the measurements of [11], since the naive discretization of the energy flow vector used in this work is in general not conserved. In view of these considerations, it is reasonable to ask whether the discrepancy between the lattice results of [11] in the deconfinement phase and the free-fermion result (2) is due to the lattice or finite-volume artifacts or due to the genuinely strong-coupling nature of the quark-gluon plasma state.

In these Proceedings we calculate the chiral vortical conductivity for free overlap fermions on the lattice and check the validity of the result (2). We use the same lattice setup as in [11] and study the naively discretized energy flow along the direction of the constant axial magnetic field.

As demonstrated in [12], overlap Dirac operator in the presence of both vector and axial gauge fields $V_\mu(x)$ and $A_\mu(x)$ should satisfy the following functional equation:

$$\frac{\partial}{\partial A_\mu(x)}\mathcal{D}_{ov}\left[V_\mu(x), A_\mu(x)\right] = \frac{\partial}{\partial V_\mu(x)}\mathcal{D}_{ov}\left[V_\mu(x), A_\mu(x)\right]\gamma_5\left(1 - \mathcal{D}_{ov}\left[V_\mu(x), A_\mu(x)\right]\right). \tag{5}$$

This equation drastically simplifies if one formulates it in terms of the overlap Dirac propagator $\mathcal{D}_{ov}^{-1}$:

$$\frac{\partial}{\partial A_\mu(x)}\mathcal{D}_{ov}^{-1} = -\mathcal{D}_{ov}^{-1}\left(\frac{\partial}{\partial A_\mu(x)}\mathcal{D}_{ov}\right)\mathcal{D}_{ov}^{-1} = -\mathcal{D}_{ov}^{-1}\left(\frac{\partial}{\partial V_\mu(x)}\mathcal{D}_{ov}\right)\gamma_5\left(1 - \mathcal{D}_{ov}\right)\mathcal{D}_{ov}^{-1} =$$
$$= \mathcal{D}_{ov}^{-1}\left(\frac{\partial}{\partial V_\mu(x)}\mathcal{D}_{ov}\right)\mathcal{D}_{ov}^{-1}\gamma_5 = -\frac{\partial}{\partial V_\mu(x)}\mathcal{D}_{ov}^{-1}\gamma_5, \tag{6}$$

where in the last line we have used the Ginsparg-Wilson relation and the arguments of $\mathcal{D}_{ov}$ were not shown for the sake of brevity. We have thus arrived at the linear equation, which has a straightforward solution in terms of the chiral projectors $\mathcal{P}_{\pm} = \frac{1\pm\gamma_5}{2}$:

$$\mathcal{D}_{ov}^{-1}\left[V_\mu, A_\mu\right] = \mathcal{P}_+ \mathcal{D}_{ov}^{-1}\left[V_\mu + A_\mu\right]\mathcal{P}_- + \mathcal{P}_- \mathcal{D}_{ov}^{-1}\left[V_\mu - A_\mu\right]\mathcal{P}_+ + 1/2, \tag{7}$$

where the propagators $\mathcal{D}_{ov}^{-1}\left[V_\mu \pm A_\mu\right]$ are obtained from the overlap Dirac operators $\mathcal{D}_{ov}\left[V_\mu \pm A_\mu\right]$ in the background of the vector gauge fields $V_\mu(x) \pm A_\mu(x)$. The latter are coupled to fermions in a standard way by including the corresponding link factors into finite difference operators entering the Dirac-Wilson operator which is used to define the overlap Dirac operator [13]. The integration constant 1/2 in (7) is fixed by the requirement that at $A_\mu(x) = 0$ one obtains the conventional overlap propagator in the background of the vector field $V_\mu(x)$ [13]. A similar construction for the overlap Dirac propagator in the presence of an axial gauge field was also used in [11], however, an integration constant of 1/2 was not taken into account. It is indeed irrelevant for the Chiral Vortical Effect. However, this constant in fact incorporates the effect of axial anomaly on the lattice [13] and can be extremely important for other anomalous transport phenomena such as Chiral Magnetic Effect or Chiral Separation Effect.

In order to implement the constant axial magnetic field on the lattice, we use the configuration of $A_\mu(x)$ which yields the constant lattice field strength $F_{xy} = -F_{yx} = A_x(x) + A_y(x+\hat{e}_x) - A_x(x+\hat{e}_y) - A_y(x)$ in (7) and set $V_\mu(x) \equiv 0$. Due to flux quantization, possible values of field strength are $F_{xy} = 2\pi N_5/L^2$, where L is the spatial lattice size and N_5 is an integer. Chiral and baryon chemical potentials are set to zero. Overlap Dirac operator is calculated and inverted using `LAPACK`. The energy flow along the axial magnetic field $J_E \equiv T_{0z}$ (4) is then calculated as in [11].

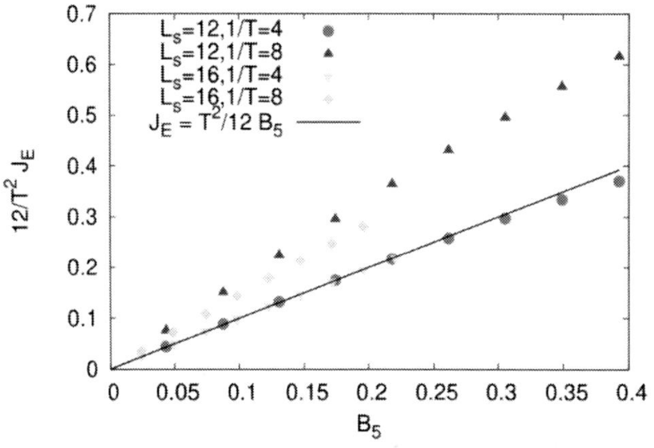

Figure 1. Energy flow (4) on the lattice as a function of axial magnetic field B_5 at different spatial lattice sizes and at different temperatures. Solid line corresponds to the continuum result (1), (2).

In Figure 1 we plot the energy flow J_E along the direction of the axial magnetic field as a function of field strength for different temperatures and different spatial lattice sizes. We assume that lattice spacing is equal to unity. One can see that in accordance with (1) the energy flow is almost a linear function of B_5 with a slight tendency for saturation at large values of B_5.

In order to extract the value of the chiral vortical conductivity, we perform the linear fits of the energy flow as a function of B_5, using data points with three smallest values of B_5. The

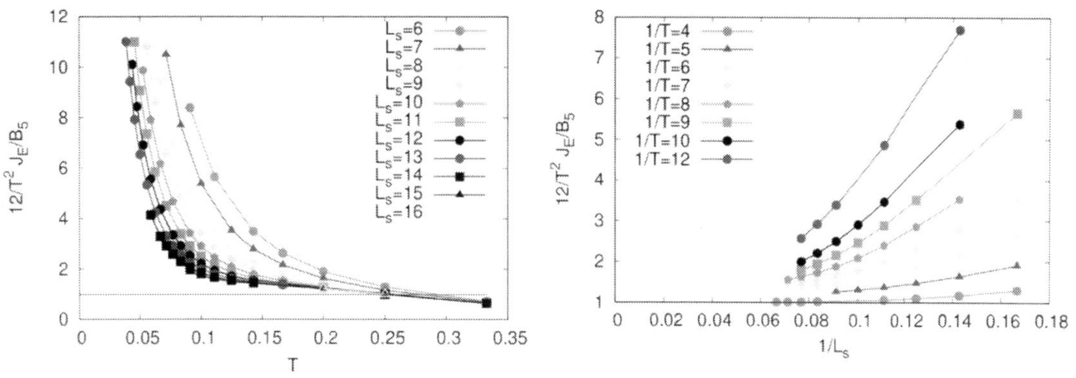

Figure 2. Finite volume scaling of the chiral vortical conductivity calculated with free overlap lattice fermions. On the left: chiral vortical conductivity as a function of temperature at different spatial lattice sizes. On the right: chiral vortical conductivity as a function of inverse lattice size at different fixed temperatures.

results are illustrated by Fig. 2 for different values of temperature and different lattice sizes. An immediate conclusion is that the finite-volume effects are very strong, several times stronger than the continuum, infinite-volume result (2). However, from the right plot on Fig. 2 one can also see that if one increases the lattice size at fixed temperature, the chiral vortical conductivity indeed tends to the continuum result (2). The smaller is the temperature, the larger should be the lattice size.

We conclude that the continuum result (2) is indeed reproduced for free overlap lattice fermions, however, finite-volume effects are very strong and one should perform a very careful extrapolation to the thermodynamic limit.

Acknowledgements

I would like to dedicate this work to the memory of my Teacher, Prof. Dr. M. I. Polikarpov. I am grateful to Dr. V. Braguta, Dr. M. Chernodub, Dr. K. Landsteiner and Dr. M. Ulybyshev for interesting discussions. This work was supported by the S. Kowalewskaja award from the Alexander von Humboldt foundation.

References

[1] Kharzeev D E, Landsteiner K, Schmitt A and Yee H 2012 Strongly interacting matter in magnetic fields: an overview in Lect. Notes Phys. "Strongly interacting matter in magnetic fields" (Springer) ArXiv:1211.6245
[2] Hongo M, Hirono Y and Hirano T 2013 First numerical simulations of anomalous hydrodynamics ArXiv:1309.2823
[3] Erdmenger J, Haack M, Kaminski M and Yarom A 2009 *JHEP* **0901** 055 ArXiv:0809.2488
[4] Banerjee N, Bhattacharya J, Bhattacharyya S, Dutta S, Loganayagam R and Surówka P 2011 *JHEP* **1101** 094 ArXiv:0809.2596
[5] Landsteiner K, Megias E and Pena-Benitez F 2011 *Phys. Rev. Lett.* **107** 021601 ArXiv:1103.5006
[6] Golkar S and Son D T 2012 (Non)-renormalization of the chiral vortical effect coefficient ArXiv:1207.5806
[7] Hou D, Liu H and Ren H 2012 *Phys. Rev. D* **86** 121703 ArXiv:1210.0969
[8] Amado I, Landsteiner K and Pena-Benitez F 2011 *JHEP* **05** 081 ArXiv:1102.4577
[9] Yamamoto A and Hirono Y 2013 *Phys. Rev. Lett.* **111** 081601 ArXiv:1303.6292
[10] Yamamoto A 2011 *Phys. Rev. Lett.* **107** 031601 ArXiv:1105.0385
[11] Braguta V, Chernodub M N, Landsteiner K, Polikarpov M I and Ulybyshev M V 2013 Numerical evidence of the axial magnetic effect ArXiv:1303.6266
[12] Buividovich P V 2013 *PoS* **Lattice2013** 179 ArXiv:1309.2850
[13] Neuberger H 1998 *Phys. Lett. B* **417** 141 ArXiv:hep-lat/9707022

Shear viscosity in $SU(2)$ lattice gluodynamics

V. V. Braguta[1,2,3], **A. Yu. Kotov**[1]

[1] Alikhanov Institute for Theoretical and Experimental Physics, ul. Bolshaya Cheremushkinskaya 25, Moscow, 117218 Russia

[2] Institute for High Energy Physics, Protvino, Moscow region, 142281 Russia

[3] Moscow Institute of Physics and Technology (State University), Institutskii per. 9, Dolgoprudnyi, Moscow region, 141700 Russia

E-mail: braguta@mail.ru, kotov@itep.ru

Abstract. The calculation of the gluon plasma viscosity has been performed by lattice simulations in the $SU(2)$ gluodynamics at $T/T_c = 1.2$ on the supercomputers. The evaluation is based on the Kubo formula that relates viscosity to the spectral function of the correlation functions of the energy-momentum tensor. For the extraction of the spectral function from the Euclidean correlator the linear method is applied.

1. Introduction

One of the most important results obtained at RHIC experiment on heavy ion collisions is the measurement of the elliptic flow of final particles[1, 2]. Obtained value can be explained within the hydrodynamical approximation if one assumes that quark-gluon plasma formed during heavy ion collisions is almost a superfluid liquid[3, 4, 5]. Numerical simulations of a relativistic liquid lead to the following bound on the ratio of the viscosity to the entropy density of the system[6]: $\eta/s \leq 0.4$, which is the smallest ratio for all systems known to date. Thereby one of the most important problems in the modern theoretical physics is a first-principle calculation of the viscosity of quark-gluon plasma. One of the most powerful approaches to this problem is the numerical simulation of quark-gluon plasma. It is worth mentioning, that there were the works, in which the viscosity of quark-gluon plasma was studied on the lattice[7, 8, 9], but due to the complicated calculations, the results obtained are not exhaustive.

In the present paper the calculation of the viscosity of the $SU(2)$ gluodynamics is performed by means of lattice modeling. Instead of the $SU(3)$ QCD we consider the $SU(2)$ model as both theories are rather similar physically and at the same time numerical calculations in the $SU(2)$ gluodynamics are much simpler than in the $SU(3)$ theory. This leads to the significant advances in the study of transport coefficients of quark-gluon plasma.

2. A brief description of the method

Kubo formulas [10] relating transport coefficients to the correlators of energy-momentum tensor are used in this calculation. In particular, for the viscosity the Kubo formula reads:

$$\eta = \pi \lim_{\omega \to 0} \frac{\rho_{12,12}(\omega, \mathbf{q} = 0)}{\omega}, \tag{1}$$

Journal of Physics: Conference Series **607** (2015) 012019 doi:10.1088/1742-6596/607/1/012019

where $\rho_{12,12}$ is a spectral function, which is determined as an imaginary part of the two-point retarded Green function: $\rho_{12,12}(\omega, \mathbf{q}) = \frac{1}{\pi} \operatorname{Im}\langle T_{12}T_{12}\rangle_{ret}(\omega, \mathbf{q})$. The Green function of energy-momentum tensor measured in lattice calculations is defined as follows:

$$C_{12,12}(x_0, \mathbf{p}) = \beta^5 \int d^3\mathbf{x} e^{i\mathbf{p}x} \langle T_{12}(0) T_{12}(x_0, \mathbf{x})\rangle, \qquad (2)$$

where $\beta = 1/T$ is an inverse temperature of the system. Function (2) is an analytical continuation of the retarded Green function: $p_0 \to i\omega$ and is related to the spectral function in the following way:

$$C_{12,12}(x_0, \mathbf{p}) = \beta^5 \int_0^\infty \rho_{12,12}(\omega, \mathbf{p}) \frac{\cosh \omega(\frac{1}{2}\beta - x_0)}{\sinh \frac{\omega\beta}{2}} d\omega \qquad (3)$$

In order to determine the value of viscosity one needs to measure on the lattice the correlator $C_{12,12}$ (1) as a function of x_0 for $\mathbf{p} = 0$, and then invert (3) to find the spectral function $\rho_{12,12}(\omega, \mathbf{q} = 0)$. Low-frequency behaviour of spectral function $\rho_{12,12}$ determines the value of the viscosity.

For the inversion of (3) we use the linear method proposed in [9], which is based on the assumption that the spectral function is rather smooth and can be decomposed into certain basis functions.

3. Measurement of the viscosity

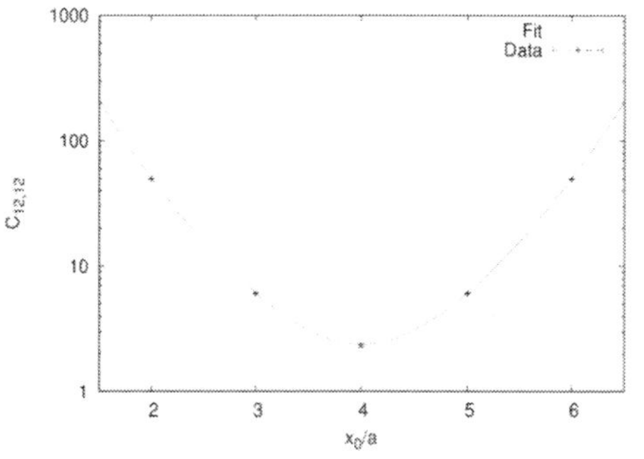

Figure 1: Correlation function $C_{12,12}(x_0)$ of the energy-momentum tensor. The points show the data measured on the lattice. The red line corresponds to the spectral function extracted from this data.

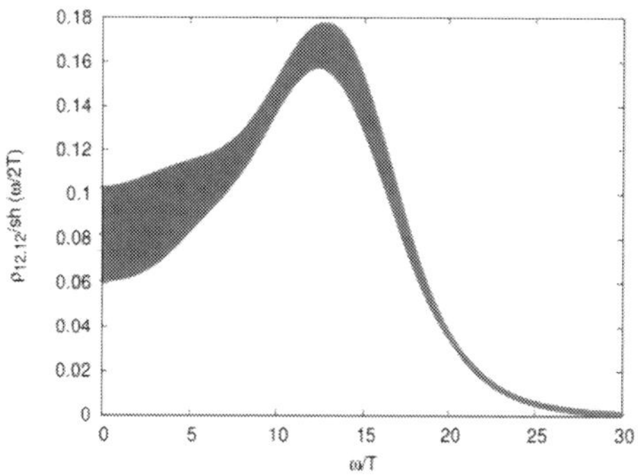

Figure 2: Spectral density $\rho_{12,12}(\omega)$ divided by $\sinh(\frac{\omega}{2T})$. The filled region corresponds to statistical uncertanties of our data.

The measurement of energy-momentum tensor was performed in lattice gluodynamics with the $SU(2)$ gauge group and the Wilson action:

$$S_g = \beta \sum_{x,\mu<\nu} \left(1 - \frac{1}{N_c} \operatorname{tr} U_{\mu,\nu}(x)\right). \qquad (4)$$

Energy-momentum tensor in gluodynamics has the form:

$$T_{\mu\nu} = F^a_{\mu\alpha} F^a_{\nu\alpha} - \frac{1}{4}\delta_{\mu\nu} F^a_{\rho\sigma} F^a_{\rho\sigma} \qquad (5)$$

For the chromo-electormagnetic tensor the clover-shaped discretization is used:

$$F_{\mu\nu}(x) = \frac{1}{4}(V_{\mu,\nu}(x) + V_{\nu,-\mu}(x) + V_{-\mu,-\nu}(x) + V_{-\nu,\mu}(x)) \qquad (6)$$
$$V_{\mu,\nu}(x) = \frac{1}{2}(U_{\mu,\nu}(x) - U_{\nu,\mu}(x)).$$

The two-level algorithm is used for measuring correlator (1). It significantly improves the accuracy of the calculations [12]. The simulations are performed on the lattice 8×32^3, $\beta = 2.643$, which corresponds to the temperature $T/T_c \approx 1.2$. The results of the evaluation of correlator (2) are shown in Fig. 1.

For the extraction of spectral function the linear method described in [9] is used. The following function that describes the behaviour of the spectral function at large frequencies (which is determined by the asymptotical freedom) is used as an initial approximation:

$$m(\omega) = \frac{A\omega^4}{\tanh^2\frac{\omega}{4T}\tanh^2\frac{\omega}{2T}} \qquad (7)$$

In Fig. 2 the obtained data for the spectral function is shown. Statistical errors are determined in accordance with the method desribed in [9].

The results for the measuring of the viscosity are presented as the ratio η/s, where s is an entropy density. The latter is determined using the relation: $s = \frac{\epsilon+p}{T}$ and the method for the evalutation of $\epsilon + p$, described in [13]. Applying this method we obtain the following value for this ratio:

$$\frac{\eta}{s} = 0.111 \pm 0.032 \qquad (8)$$

An error in this result corresponds to the statistical uncertainty of our data. Value (8) is close to the so-called KSS bound $\eta/s \geq 1/4\pi \approx 0.08$, obtained by means of AdS/CFT-correspondence [14]. Moreover, result (8) satisfies the bound obtained from the description of RHIC experiments $\eta/s < 0.4$ [6].

Acknowledgments

The authors wish to thank V. G. Bornyakov and M. I. Polykarpov for interesting and useful discussions. The calculations were performed at ITEP clusters "Stakan" and "Graphin"(the authors are grateful to A. V. Barilov, A. A. Golubev, V. A. Kolosov, I. E. Korol'ko and M. M. Sokolov for assistance). The work was supported by RFBR grant 11-02-01227-a. A. K. acknowledges the support of the Dynasty foundation.

References

[1] Adams J *et al* [STAR Collaboration] 2005 *Nucl. Phys.* A **702** 102 (*Preprint* nucl-ex/0501009)
[2] Adcox K *et al* [PHENIX Collaboration] 2005 *Nucl. Phys.* A **757** 184 (*Preprint* nucl-ex/0410003)
[3] Kolb P F, Huovinen P, Heinz U W and Heiselberg H 2001 *Phys. Lett.* B **500** 232 (*Preprint* hep-ph/0012137)
[4] Huovinen P, Kolb P F, Heinz U W, Ruuskanen P V and Voloshin S 2001 *Phys. Lett.* B **503** 58 (*Preprint* hep-ph/0101136)
[5] Teaney D, Lauret J and Shuryak E V 2001 *Phys. Rev. Lett.* **86** 4783 (*Preprint* nucl-th/0011058)
[6] Teaney D 2009 Viscous Hydrodynamics and the Quark Gluon Plasma *Preprint* 0905.2433
[7] Nakamura A and Sakai S 2005 Transport coefficients of Gluon Plasma *Phys. Rev. Latt.* **94** 072305 (*Preprint* hep-lat/0406009)
[8] Karsch F and Wyld H W 1987 Thermal Green's functions and transport coefficients on the lattice *Phys. Rev.* D **35** 2518

[9] Meyer H B 2007 A Calculation of the shear viscosity in SU(3) gluodynamics *Phys. Rev.* D **76** 101701 (*Preprint* 0704.1801)
[10] Kubo R 1957 Statistical mechanical theory of irreversible processes. 1. General theory and simple applications in magnetic and conduction problems *J. Phys. Sox. Jap.* **12** 570
[11] Meyer H B 2011 Transport Properties of the Quark-Gluon Plasma *Eur. Phys. J.* A **47** 86 (*Preprint* 1104.3708)
[12] Meyer H B 2003 Locality and statistical error reduction on correlation functions *J. High Energy Phys.* JHEP01(2003)048 (*Preprint* hep-lat/0209145)
[13] Engels J, Karsch F and Scheideler T 2000 *Nucl. Phys.* B **564** 303 (*Preprint* hep-lat/9905002)
[14] Kovtun P, Son D T and Starinets A O *Phys. Rev. Lett.* **94** 111601 (*Preprint* hep-ph/0203177)

Emergent gravity in graphene

dedicated to the memory of M.I.Polikarpov

M.A. Zubkov
University of Western Ontario, Department of Applied Mathematics, London, ON, Canada N6A 5B7;
ITEP, B.Cheremushkinskaya 25, Moscow, 117259, Russia

G.E. Volovik
Low Temperature Laboratory, School of Science and Technology, Aalto University, P.O. Box 15100, FI-00076 AALTO, Finland;
L. D. Landau Institute for Theoretical Physics, Kosygina 2, 119334 Moscow, Russia

Abstract. We reconsider monolayer graphene in the presence of elastic deformations. It is described by the tight - binding model with varying hopping parameters. We demonstrate, that the fermionic quasiparticles propagate in the emergent $2D$ Weitzenbock geometry and in the presence of the emergent $U(1)$ gauge field. Both emergent geometry and the gauge field are defined by the elastic deformation of graphene.

1. Introduction

It can be shown [1, 2, 3, 4, 5, 6], that in 3 + 1 D systems with fermions near the Fermi - points the two - component spinors ψ appear. The remaining components of the original fermion field are in general case massive, and, therefore, decouple in the low energy limit. These Weil spinors are described by action

$$\mathcal{S} = \int d^D x |\det e| \, e^{\mu}_a \, \bar{\psi} \sigma^a (p_\mu - \mathcal{A}_\mu) \psi + \dots \tag{1.1}$$

In this action the collective mode $\mathcal{A}_\mu$ may contain the shift of the Fermi point resulted in the effective gauge field A_μ and the effective spin connection $\frac{i}{8} C^{ab}_\mu [\sigma^a, \sigma^b]$. The modes e^μ_a may be considered as the vielbein describing the gravitational degrees of freedom[1]. Here we demonstrate, that the same phenomenon occurs in the 2 + 1 D model of monolayer graphene. The corresponding field $\mathcal{A}_\mu$ contains the $U(1)$ gauge field and does not contain any spin connection (in the leading approximation in elastic deformations). The corresponding mode e^μ_a contains the nontrivial $2D$ zweibein. Both these emergent fields are expressed through strain.

[1] It is worth mentioning that there is the conceptually different approach to emergent gravity, in which the bilinear combinations $\psi \hat{O}^a_i \psi$ or $\bar{\psi} \hat{O}^a_i \psi$ are identified with the vielbein for some tensorial operators $\hat{O}$. Corresponding constructions were considered in the context of relativistic field theory in [7, 8] and in the context of condensed matter physics in [9].

2. The tight - binding model with varying hopping parameters. Floating Fermi - point as the emergent gauge field.

The carbon atoms of graphene form a honeycomb lattice with two sublattices A and B (of the triangular form). We denote the lattice spacing by a. Let us introduce vectors that connect a vertex of the sublattice A to its neighbors (that belong to the sublattice B): $\mathbf{l}_1 = (-a, 0)$, $\mathbf{l}_2 = (a/2, a\sqrt{3}/2)$, $\mathbf{l}_3 = (a/2, -a\sqrt{3}/2)$.

We suppose that the hopping parameter varies, so that its value depends on the particular link connecting two adjacent points of the honeycomb lattice. This is caused by elastic deformations. We have three values of $t_a, a = 1, 2, 3$ at each point. The Hamiltonian has the form

$$H = -\sum_{\alpha \in A}\sum_{j=1}^{3} t_j(\mathbf{r}_\alpha)\Big(\psi^\dagger(\mathbf{r}_\alpha)\psi(\mathbf{r}_\alpha + \mathbf{l}_j) + \psi^\dagger(\mathbf{r}_\alpha + \mathbf{l}_j)\psi(\mathbf{r}_\alpha)\Big), \tag{2.1}$$

In addition, we define the following variables: $\mathbf{m}_1 = -\mathbf{l}_1 + \mathbf{l}_2$, $\mathbf{m}_3 = -\mathbf{l}_3 + \mathbf{l}_1$, $\mathbf{m}_2 = -\mathbf{l}_2 + \mathbf{l}_3 = -\mathbf{m}_1 - \mathbf{m}_3$. In momentum space the effective Hamiltonian has has the form $H = \int \frac{d^2k}{\Omega}\frac{d^2k'}{\Omega}\psi^\dagger(\mathbf{k}')\,\hat{U}\Big(t_j(\mathbf{k}_-), \mathbf{k}_+, \mathbf{k}_-\Big)\,\psi(\mathbf{k})$, where

$$\hat{U}(t_j, \mathbf{k}_+, \mathbf{k}_-) = -\sum_{j=1}^{3} t_j \begin{pmatrix} 0 & e^{-i\mathbf{l}_j(\mathbf{k}_+ + \mathbf{k}_-)} \\ e^{i\mathbf{l}_j(\mathbf{k}_+ - \mathbf{k}_-)} & 0 \end{pmatrix} \tag{2.2}$$

while Ω is the area of momentum space, $\mathbf{k}_+ = (\mathbf{k} + \mathbf{k}')/2$ and $\mathbf{k}_- = (\mathbf{k} - \mathbf{k}')/2$. We imply, that the variations of t_a are given by

$$t_a(\mathbf{r}) = t(1 - \Delta_a(\mathbf{r})), \quad |\Delta_a| \ll 1 \tag{2.3}$$

We introduce the notation $K^\pm(t_j)$ for the "floating Fermi point". It is defined in such a way, that for $\mathbf{k}_+ = K^\pm(t_j)$, $\mathbf{k}_- = 0$ the function $\hat{U}(t_j, \mathbf{k}_+, \mathbf{k}_-)$ vanishes: $\hat{U}(t_j, K^\pm(t_j), 0) = 0$. For the case of homogenious variations of hopping parameters when $\mathbf{k}_- = 0$ this definition gives the true Fermi point. For $\mathbf{k}_- \neq 0$ the interpretation is not so obvious. In general case the eigenvectors of the one - particle Hamiltonian do not correspond to the definite value of momentum. We have the complicated wave - packets instead. However, in the next section it will be shown that in the linear approximation of $\hat{U}(t_j, \mathbf{k}_+, \mathbf{k}_-)$ (as a function of $\mathbf{k}_+$, $\mathbf{k}_-$) the term proportional to $\mathbf{k}_-$ may be neglected in the considered theory. This brings us back to the interpretation of $K^\pm(t_j)$ as the Fermi point. Now we take into account that hopping parameters t_j themselves depend on $\mathbf{k}_-$. That's why $K^\pm(t_j(\mathbf{k}_-))$ is the momentum - dependent ("floating") Fermi point. One can easily find $K^\pm \approx K^{(0)}_\pm \pm \mathbf{A}$, where in addition to the fixed Fermi - point $K^{(0)}_\pm = \pm\frac{4\pi}{a^2 3\sqrt{3}}\Big(-\frac{\mathbf{m}_2}{\sqrt{3}}\Big)$ the emergent $U(1)$ gauge field $\mathbf{A}$ appears with the components [11]:

$$\mathbf{A}^b = -\frac{2}{3a^2}\epsilon^{ba}\sum_j \Delta_j \mathbf{l}^a_j \tag{2.4}$$

3. Expansion near the floating fermi - point

Next, we expand $\hat{U}(t_j, \mathbf{k}_+, \mathbf{k}_-)$ around $\mathbf{k}_+ = K^\pm(t_j)$, $\mathbf{k}_- = 0$. The result has the form:

$$\hat{U}(t_j, \mathbf{k}_+, \mathbf{k}_-) = -i\sigma^3\Big[(\mp\sigma^2\mathbf{f}_2 + \sigma^1\mathbf{f}_1)\Big(\mathbf{k}_+ - K^\pm(t_j)\Big) - (\sigma^1\mathbf{f}_1 \mp \sigma^2\mathbf{f}_2)\sigma^3\mathbf{k}_-\Big], \tag{3.1}$$

where $\mathbf{f}$ (as well as $\mathbf{A}$) depends on $\mathbf{k} - \mathbf{k}'$ and is given by

$$\mathbf{f}^k_a = v_F\Big(\delta^k_a - \frac{2}{3a^2}\sum_j \Delta_j\Big[\mathbf{l}^a_j\mathbf{l}^k_j - \frac{a}{2}\mathbf{l}^d_j K^{dak}\Big]\Big) \tag{3.2}$$

Here we introduced [10] the new tensor K: $K^{ijk} = -\frac{4}{3a^3}\sum_b \mathbf{l}_b^i\mathbf{l}_b^j\mathbf{l}_b^k$, $K^{111} = -K^{122} = -K^{221} = -K^{212} = 1$. One can see, that the emergent $U(1)$ field is related to the field $\mathbf{f}$ as follows:

$$\mathbf{A}^i = -\frac{1}{2v_F a}\epsilon^{ik}K^{kjb}\mathbf{f}_b^j \tag{3.3}$$

We define the new spinors: $\Psi_\pm(\mathbf{Q}) = \psi(K_\pm^{(0)} + \mathbf{Q})$. In order to return to the coordinate space we should use the following rule $\frac{1}{2}(\mathbf{Q}+\mathbf{Q}')F(\mathbf{Q}-\mathbf{Q}') \to -\frac{i}{2}(F(\mathbf{x})\overrightarrow{\nabla} - \overleftarrow{\nabla}F(\mathbf{x}))$, $\frac{1}{2}(\mathbf{Q}-\mathbf{Q}')F(\mathbf{Q}-\mathbf{Q}') \to -\frac{i}{2}\Big(\nabla F(\mathbf{x})\Big)$. As a result the hamiltonian has the form $H = \sum_\pm \int d^2x[\Psi^\pm(\mathbf{x})]^\dagger \mathbf{H}_\pm \Psi^\pm(\mathbf{x})$, where

$$\mathbf{H}_- = -\sigma^3\,\mathbf{f}_a^k(\mathbf{x})\sigma^a \circ [\partial_k + i(\mathbf{A}_k(\mathbf{x}) + \tilde{\mathbf{A}}_k(\mathbf{x}))];\ \mathbf{H}_+ = -\sigma^2\Big(\sigma^3\,\mathbf{f}_a^k(\mathbf{x})\sigma^a \circ [\partial_k - i(\mathbf{A}_k(\mathbf{x}) + \tilde{\mathbf{A}}_k)]\Big)\sigma^2$$

Here the field $\mathbf{f}$ is defined in coordinate space and is related to the variables Δ_a according to Eq. (3.2). The product $i\mathbf{f}_a^k \circ \partial_k$ in these equations should be understood as $\mathbf{f}_a^k \circ i\partial_k = \frac{i}{2}\Big(\mathbf{f}_a^k\overrightarrow{\partial_k} - \overleftarrow{\partial_k}\mathbf{f}_a^k\Big)$. The additional gauge field $\tilde{\mathbf{A}}$ appears: $\tilde{\mathbf{A}}_a(\mathbf{x}) = \frac{1}{2v_F}\nabla_i \mathbf{f}_b^i(\mathbf{x})\epsilon_{ba}$. This additional field $\tilde{\mathbf{A}}$ is to be compared with the emergent gauge field $\mathbf{A}$. One can see that $\tilde{\mathbf{A}} \sim a\nabla\mathbf{A}$. Therefore, this is not reasonable to keep this additional field together with $\mathbf{A}$ in the field - theoretical description, where all dimensional quantities are to be much larger than the lattice spacing a. This shows that even in case of the variations of t_a depending on the position in coordinate space we may omit the derivatives of $t_a(x)$ in the effective low energy field - theoretical Hamiltonian (that is we may omit the terms $\sim \mathbf{k}_-$ in momentum space representation). The field $\mathbf{f}$ is related to the dreibein $\mathbf{e}$ as follows: $e\,\mathbf{e}_a^i = \mathbf{f}_a^i$; $e\,\mathbf{e}_0^0 = 1$, $\mathbf{e}_a^0 = \mathbf{e}_0^i = 0$, where $i, a = 1,2$. Here the determinant of the zweibein $\det \mathbf{e}_a^k = 1$, where $a,k = 1,2$. At the same time the three - dimensional determinant of $\mathbf{e}$ is equal to $\det \mathbf{e}^{(3\times 3)} = \mathbf{e}_0^0 = 1/e$. The three - volume element is $d^{(3)}V = d^2\mathbf{r}dte(\mathbf{r},t)$. The value of e is given by $e = [\det \mathbf{f}]^{1/2} = v_F(1 - \frac{1}{3}(\Delta_2 + \Delta_3 + \Delta_1))$.

4. Elastic deformations as a source of emergent gravity

The graphene sheet is parametrized by variable $x_k, k = 1,2$. The classical elasticity theory has the displacements $u_a(x)$ as degrees of freedom ($a = 1,2,3$). The three - dimensional coordinates y_a of the graphene sheet are given by

$$y_k(x) = x_k + u_k(x), \quad k = 1,2; \quad y_3(x) = u_3(x) \tag{4.1}$$

At $u_a = 0$ the graphene is flat. The emergent metric of elasticity theory is given by $g_{ik} = \delta_{ik} + 2u_{ik}$, $u_{ik} = \frac{1}{2}\Big(\partial_i u_k + \partial_k u_i + \partial_i u_a \partial_k u_a\Big)$, $a = 1,2,3$, $i,k = 1,2$. According to [10] the elastic deformations of graphene result in the change of the hopping elements, which determine the effective geometry experienced by fermions. The simplest connection between the deformations and the hopping elements t_n ($n = 1,2,3$), which is allowed by symmetry, is

$$t_a(\mathbf{r}) = t[1 - \frac{\beta}{a^2}u_{ik}(\mathbf{r})\mathbf{l}_a^i\mathbf{l}_a^k]\,. \tag{4.2}$$

The dimensionless phenomenological parameter β is determined by the microscopic physics. As it was mentioned above, the given consideration works for the displacements u_a that are not necessarily small. However, we imply that $\beta|u_{ij}| \ll 1$. This is the requirement that the derivatives of u_a are small being multiplied by β.

The emergent geometry and emergent $U(1)$ gauge field are given by $\mathbf{e}_a^i = \mathbf{f}_a^i/\Big(\det\mathbf{f}\Big)^{1/2}$ and Eq. (3.3) with $\mathbf{f}_a^i = v_F\left(\delta_a^i - \beta\begin{bmatrix} u_{11} & u_{21} \\ u_{12} & u_{22}\end{bmatrix}\right)$. This results in the usual expression for the strain

- induced electromagnetic field (see [10, 12, 13, 14, 15] and references therein). As for our values of $\mathbf{f}^k_a$, they differ from the expression for the anisotropic Fermi velocity calculated in [10]. The zweibein is given by $\mathbf{e}^i_a = \mathbf{f}^i_a/e$. It is constructed in such a way that the determinant of the zweibein $\det \mathbf{e}^{(2\times 2)} = 1$. The 2 + 1 D volume element $d^{(3)}V = e(\mathbf{r}, t)\, d^2\mathbf{r}dt$ corresponds to the function $e = v_F(1 - \frac{\beta}{2}u_{aa})$. This function is related to the $(^0_0)$ - component of the dreibein as $\mathbf{e}^0_0 = 1/e$. The determinant of the dreibein is equal to $1/e$.

5. Conclusions

Here we considered the long - standing problem about the type of the geometry experienced by fermionic quasiparticles in graphene in the presence of elastic deformations. The appearance of such a geometry allows to model gravity within graphene [16]. In some of the previous works on this subject it was supposed that such an emergent geometry is Riemannian [13]. Later, the derivation of the space - dependent Fermi velocity in the presence of strain was undertaken [10]. Here we present the direct derivation of the emergent geometry. In our approach the expansion of the effective hamiltonian is performed near to the "floating" Fermi point (where the hamiltonian vanishes in the limit of homogenious elastic deformations). We demonstrate, that the emergent geometry is described by the $2D$ zweibein. Also, the emergent $U(1)$ gauge field appears (its expression through strain coincides with the one derived previously [11]) while the spin connection does not appear in the approximation linear in elastic deformations[2]. The emergent $U(1)$ field, in turn, is expressed through the zweibein. Thus we deal with the varying $2D$ Weitzenbock geometry expressed through the elastic deformation. This means that there is the emergent teleparallel gravity.

The authors kindly acknowledge discussions of emergent gravity with D.I.Diakonov, and useful correspondence with M.Vozmediano. This work was partly supported by RFBR grant 11-02-01227, by the Federal Special-Purpose Programme 'Human Capital' of the Russian Ministry of Science and Education. The work of M.A.Z. is partly supported by the Natural Sciences and Engineering Research Council of Canada. GEV acknowledges a financial support of the Academy of Finland and its COE program, and the EU FP7 program (#228464 Microkelvin).

[1] P. Hořava, Stability of Fermi surfaces and K-theory, Phys. Rev. Lett. **95**, 016405 (2005).
[2] H.B. Nielsen, M. Ninomiya: Absence of neutrinos on a lattice. I - Proof by homotopy theory, Nucl. Phys. B **185**, 20 (1981); Absence of neutrinos on a lattice. II - Intuitive homotopy proof, Nucl. Phys. B **193**, 173 (1981).
[3] C.D. Froggatt and H.B. Nielsen, *Origin of Symmetry*, World Scientific, Singapore, 1991.
[4] G.E. Volovik, The Universe in a Helium Droplet, Clarendon Press, Oxford (2003).
[5] G.E. Volovik, Analog of gravity in superfluid ^{3}He-A, JETP Lett. **44**, 498–501 (1986).
[6] G.E. Volovik, Topology of quantum vacuum, Chapter in proceedings of the Como Summer School on analogue gravity, arXiv:1111.4627.
[7] C. Wetterich, Gravity from spinors, Phys. Rev. D **70**, 105004 (2004); Universality of geometry, Phys. Lett. B **712**, 126–131 (2012).
[8] D. Diakonov, A.G. Tumanov, A.A. Vladimirov, Low-energy General Relativity with torsion: a systematic derivative expansion, Phys. Rev. D **84**, 124042 (2011); D. Diakonov, Towards lattice-regularized quantum gravity, arXiv:1109.0091.
[9] G.E. Volovik, Superfluid ^{3}He–B and gravity, Physica B **162**, 222 (1990).
[10] Fernando de Juan, Mauricio Sturla, and Maria A.H. Vozmediano, Phys. Rev Lett. 108, 227205 (2012);
[11] Ken-ichi Sasaki, Yoshiyuki Kawazoe, and Riichiro Saito, Local Energy Gap in Deformed Carbon Nanotubes, Progress of Theoretical Physics, Vol. 113, No. 3, March 2005
[12] F. de Juan, A. Cortijo and M.A.H. Vozmediano, Dislocations and torsion in graphene and related systems, Nucl. Phys. B **828**, 625–637 (2010).
[13] Maria A. H. Vozmediano, Fernando de Juan and Alberto Cortijo, Journal of Physics: Conference Series 129 (2008) 012001 doi:10.1088/1742-6596/129/1/012001
[14] M. A. H. Vozmediano, M. I. Katsnelson, and F. Guinea, Phys. Rep. 496 109(2010).
[15] Fernando de Juan, Juan L. Manes, and Maria A.H. Vozmediano, Phys.Rev. B 87 , 165131 (2013)
[16] D.V. Khveshchenko, "Simulating holographic correspondence in flexible graphene", arXiv:1305.6651

[2] Since the spin connection is not forbidden by symmetry, it should appear in the next approximations.